Tribology and Applications of Self-Lubricating Materials

Tribology and Applications of Self-Lubricating Materials

by

Emad Omrani

Pradeep K. Rohatgi

Pradeep L. Menezes

CRC Press
Taylor & Francis Group
Boca Raton London New York

CRC Press is an imprint of the
Taylor & Francis Group, an **informa** business

CRC Press
Taylor & Francis Group
6000 Broken Sound Parkway NW, Suite 300
Boca Raton, FL 33487-2742

First issued in paperback 2019

ISBN-13: 978-1-4987-6848-1 (hbk)
ISBN-13: 978-0-367-87826-9 (pbk)

Library of Congress Cataloging-in-Publication Data

Names: Omrani, Emad, author. | Rohatgi, P. K., author. | Menezes, Pradeep L., author.
Title: Tribology and applications of self-lubricating materials / Emad Omrani, Pradeep K. Rohatgi, and Pradeep L. Menezes.
Description: Boca Raton : CRC Press, Taylor & Francis, 2017. | Includes bibliographical references.
Identifiers: LCCN 2017026592 | ISBN 9781498768481 (hardback : acid-free paper) | ISBN 9781315154077 (ebook)
Subjects: LCSH: Tribology. | Lubrication and lubricants.
Classification: LCC TJ1075 .O47 2017 | DDC 621.8/9--dc23
LC record available at https://lccn.loc.gov/2017026592

Visit the Taylor & Francis Web site at
http://www.taylorandfrancis.com

and the CRC Press Web site at
http://www.crcpress.com

Contents

Preface

Self-lubricating materials are seen to be an integral part of today's engineering materials, which are replacing conventional materials used in automobiles, aerospace, and marine applications. The characteristic multitasking ability of these materials has made them vital for tribological and mechanical applications. The complex mechanisms behind self-lubricating composites are a cumulation of interdisciplinary concepts centered around the tribological nature of the materials and its functionality. The key elements involve understanding the compatible nature of different composites with various lubricating mechanisms along with necessary mechanical properties. Accomplishing balance of these aspects is what defines the functionality of a self-lubricating composite. This book provides an insight into the intricacies of designing self-lubricating composite materials based on the widely used engineered composite materials, such as metal, polymer, and ceramic matrix composites.

This book emphasizes on the developmental processes of various composite materials, which are used along with a specific form of lubricating mechanism. This compatibility of composites with lubricants is found to be the driving force behind the development of various self-lubricating materials. Further, efficient simplistic functioning and usability of these self-lubricating materials have fueled the research on various self-lubricating mechanisms. This book addresses these developments in detail for specific engineered composites along with providing the basic understanding of the self-lubricating materials for a wide variety of applications.

Chapter 1 details the technological and tribological needs that led to the development of self-lubricating materials. The role of solid lubricants, powder metallurgy, and film-coating techniques is found to be very critical in the development of a new class of self-lubricating materials. Further, this chapter details the advancement of self-lubricating technology and its climb from a mere bearing application to extreme environment applications. The techniques adopted over the decades to analyze the performance of these materials are also listed along with their standardized versions of testing that are currently in existence.

Chapter 2 centers around widely used self-lubricating metal matrix composites (SLMMCs) such as aluminum, copper, magnesium, and nickel matrix composites and their compatibility with solid lubricants that imbibe majority of the MMCs with self-lubricating nature. Further, this chapter addresses the tribology of carbon-based and $MoS_2/hBN/WS_2/CaF_2/BaF_2$ SLMMCs with special attention to the role of third bodies in determining the performance of these materials.

Chapter 3 addresses the nature of various self-lubricating polymer matrix composites such as epoxy, polytetrafluoroethylene (PTFE), polyetheretherketone (PEEK), phenolic, polyamide, and polystyrene. The polymer matrix composites are found to have solid lubricant as a coating on the tribological surface or inside a composite material that provide for their self-lubricating nature.

Chapter 4 details the results of a systematic study conducted to evaluate the feasibility of achieving a low friction coefficient in self-lubricating ceramic matrix composites. In addition, the feasibility of lubricating different ceramic matrix with intercalated several solid lubricants is also discussed.

Chapter 5 deals with molecular dynamic simulation of self-lubricating composites, which have been found to be an effective method to simulate and model the nano to micro level tribological properties. This chapter puts forth the importance of molecular dynamic simulations for self-lubricating composite materials.

Early career track professionals and university students need the concise information of various disciplines under the broader area of surface science, tribology, lubrication science, and novel self-lubricating materials, which this book will precisely provide. This book showcases latest knowledge and technologies available in the field that will benefit targeted reader from education, engineering industry, and scientific disciplines including but not limited to mechanical, materials, manufacturing, automotive, aerospace, and chemical.

The authors of this book are from the education industry and are closely working on industry problems. This book will find a place in middle way between the reference resources such as encyclopedia or handbooks and the regular research books. Due to the unique design of each chapter, this book serves as a good resource until there is a major breakthrough in the field of self-lubricating materials.

This book has been possible with the combine efforts of various research groups and the authors would acknowledge the research groups of Dr. Pradeep L. Menezes at the University of Nevada, Reno, Nevada and Dr. Pradeep K. Rohatgi at University of Wisconsin–Milwaukee.

Emad Omrani
University of Wisconsin–Milwaukee

Pradeep K. Rohatgi
University of Wisconsin–Milwaukee

Pradeep L. Menezes
University of Nevada, Reno

Authors

Emad Omrani is a PhD candidate at the Department of Materials Science and Engineering, Center for Advanced Materials Manufacturing, University of Wisconsin–Milwaukee (UWM), Wisconsin. His areas of research are development of novel composite materials and lightweight alloys, wear and tribology, carbounous materials as oil additive, and self-lubricating systems. He has coauthored 2 book chapters and more than 20 peer-reviewed scientific papers on metal matrix micro- and nanocomposites, biopolymeric composites, functional autonomous materials, and tribology of nanocomposites.

Pradeep K. Rohatgi received his undergraduate degree from Indian Institute of Technology (BHU), Varanasi, India and his doctorate in science from Massachusetts Institute of Technology (MIT), Cambridge in 1964. After studying at MIT, he served as a professor at the Indian Institute of Science (IISc), Bangalore, India and Indian Institute of Technology (IIT) Kanpur, India. He also served as the founder, director, and chief executive of two CSIR national laboratories including National Institute of Interdisciplinary Research, Thiruvananthapuram, India and Advanced Materials and Processing Research Institute, Bhopal, India. He currently serves as a UWM distinguished professor and director of the UWM Centers for Composites and Advanced Materials Manufacture. He has coedited and coauthored 12 books and more than 400 scientific papers and has 19 U.S. patents. He is considered as a world leader in composites and materials policy for the developing world. He has received numerous awards worldwide for his excellence in research including The Minerals, Metals & Materials Society (TMS) Chalmers Award and the American Society of Mechanical Engineers (ASME) Tribology Award and has been selected for fellowships of several organizations including TMS, American Society for Metals (ASM), ASME, Society of Automotive Engineers (SAE), The World Academy of Sciences (TWAS), Society of Manufacturing Engineers (SME), the American Association for the Advancement of Science (AAAS), Materials Research Society (MRS), and the Wisconsin Academy. His initial research on cast metal composites has been listed as a major landmark in the 11,000-year history of metal casting, and TMS organized the Rohatgi Honorary Symposium to honor his contributions to metal–matrix composites in 2006. He has developed several lightweight composite materials to reduce energy consumption. He has served on committees of governments of the United States and India in the areas of materials, especially those related to automotive sector, to promote collaboration. Currently he is working on advanced manufacture of lightweight, energy absorbing, self-lubricating, and self-healing materials and components including micro- and nanocomposites and syntactic foams. He has been a consultant to major corporations through his consulting company Future Science and Technology LLC, United Nations and World Bank and other international development agencies. He is the founder chief technology officer (CTO) of Intelligent Composites, LLC, Milwaukee, Wisconsin.

Pradeep L. Menezes (Corresponding Author) is an assistant professor in the Department of Mechanical Engineering at the University of Nevada Reno (UNR), Reno, Nevada. Before joining this university, he worked as an adjunct assistant professor at the University of Wisconsin–Milwaukee, Wisconsin and as a research assistant professor at the University of Pittsburgh, Pennsylvania. Dr. Menezes completed his PhD in materials engineering in 2008 from the Indian Institute of Science in Bangalore, India. Afterward, he spent 7 years as a postdoctoral researcher gaining experience from the University of Pittsburgh and the University of Wisconsin–Milwaukee. During the past 15 years of active research, he has gained experience in the field of materials, mechanical, and manufacturing engineering. His productive research career has produced more than 60 peer-reviewed journal publications (citations more than 1700, h-index–23), 20 book chapters, and a book titled *Tribology for Scientists and Engineers*. He has supervised more than 50 undergraduate, graduate, and postgraduate students for their research projects and completion of their degrees. Dr. Menezes serves as a reviewer for more than 50 prestigious journals and as an editorial board member of three journals. In addition, he has participated in many national and international conferences in roles such as conference paper reviewer, conference review committee member, conference technical committee member, and session chairman. He has served as the convener of the Tribology Consortium at the University of Wisconsin–Milwaukee from 2009 to 2015. His research interests include experimental and computational analysis in advanced green and bio-manufacturing, green solid and liquid lubricants and multifunctional biobased hybrid lubricants, surface science and coatings, shoe-floor design and human tribology, triboluminescence, additive manufacturing, sustainable, self-healing and self-lubricating composite materials, tribology of manufacturing systems, rock-drilling technology, and explicit finite element modeling.

1 Self-Lubricating Materials

1.1 INTRODUCTION

The need for lubrication arises when the tribological applications in mechanical systems become severe and challenging to control friction and wear. Widely used crude oil-based lubricants such as those used in automobiles (transmission, hydraulic fluids, and gear oils), metalworking fluids, alkanes, alkylbenzenes, mineral spirits, heptanes, hexanes, and other isomers are diverse enough to encompass many industrial applications. However, most of these oils on their own are limited by working conditions, physical and chemical characteristics, efficiency, and durability over extended periods of usage in severe environments. These limitations and effectiveness have been well established decades ago in the late 1960s, and as new petroleum-based oil forms are being formulated, there are still an ongoing investigation [1–4]. Considering some extreme environment applications such as optical and thermal control surfaces on spacecraft, oils and greases tend to migrate, volatize, and condense under these conditions. Even in aircraft engines, lubricants evaporate, oxidize, and decompose at high temperatures and altitudes. All these features of crude-based oils and lubricants if applied to today's high-performance machineries, automobile and turbine engines, and other advanced mechanical systems tend to limit the expectancy and efficiency of the mechanical system as a whole.

Around the same time in the late 1970s, when limitations of crude-based oils and lubricants were being established, researches understood that these oils in themselves are not sufficient enough to aid severe and advanced tribological applications. This led to the need for solid lubricants such as boric acid (H_3BO_3) and hexagonal boron nitride (hBN), which showed potential to replace oils and greases for such demanding applications. The initial idea behind using solid lubricants was that they could be applied as a coating, which not just lubricated the surface but also increased the wear resistance to extend the lifetime of the mechanical components. Based on their formulation, they were also used to strongly reduce friction and save energy. A general design requirement for a coated surface in contact was presented by Holmberg et al. [5] in the early twentieth century as follows:

- The initial coefficient of friction (COF), the steady-state COF, and the friction instability must not exceed certain design values.
- The wear of the contacting surfaces, including the coated one, must not exceed certain design values.
- The lifetime of the system must, with a specified probability, be longer than the required lifetime. The lifetime limit of the system may be defined as the time when even one of the earlier requirements is not maintained.

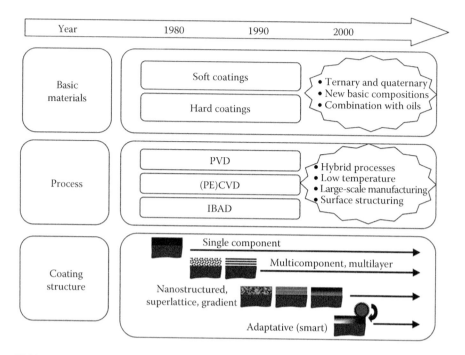

FIGURE 1.1 Historical development of tribological coatings and solid lubricant films. (Donnet, C. and Erdemir, A., *Surf. Coat. Technol.*, 180–181, 76–84, 2004.)

There have been persistent downsides of solid lubricants as well, which include poor thermal conductivity at interface, fluctuating COF under varying working conditions, and finite wear life. Compared to liquid lubricants, they are difficult to replenish; oxidation, structural chemistry changes, and aging-related degradations may also occur. Over the past few decades, most of these limitations have led to improved feasibility of solid lubricants for piratical engineering applications. A brief schematic of historical developments is shown in Figure 1.1.

The early investigations into dry bearings were based on the factors affecting the base material performances (steel, bronze, cobalt alloys) using solid lubricant films such as polytetrafluoroethylene (PTFE) + Pb and MoS_2 + polyimide. During these investigations, the focus was more on the base material rather than on the solid lubricants, but this resulted in better understanding of the latter. One of the major conclusions of this investigation was that the MoS_2-based solid lubricant resulted in significantly low friction values along with improved wear resistance properties of the coated surface [7,8]. This was followed by the National Aeronautics and Space Administration (NASA)'s own investigation into the tribological properties of MoS_2 films, where an optimum thickness range of the coating was found for which the solid lubricant was at its peak performance [9]. Even though these researchers have not yet been concerned with prioritizing additives, they still mark the beginning of a new age of lubricants whose properties could be controlled and customized considerably.

Over the past few decades, there has always been a debate over the advantages of using solid lubricants as a coating or as dispersed particles in oil. The understanding of this concept has been neither easy nor simple. As with any debate, there has been a controversial discussion as to which lubricating mechanism is more efficient [10,11]. There have been early experiments with solid lubricants such as hBN and H_3BO_3 that were mixed with lubricating oils and tested under high-pressure and high-temperature conditions [12,13]. There are several hypotheses on the mechanism of boundary lubrication, but the issue concerning the usage of different forms of solid lubricants is still an ongoing research debate [14–17]. It is safe to suggest that the end application usually dictates the way in which these solid lubricants are to be utilized, for example, in humid versus dry applications and operating temperatures of the solid lubricants.

Solid lubricants and self-lubricating properties of material go hand in hand. This is to say that it is the coatings or a composite combination of the material with these solid lubricants that imbibe self-lubricating properties to a material. The lubricants themselves may be separated into two general groups: (1) non-replenishable films, mainly applied as thin, sacrificial layers ranging from the order of 10^{-10} m to as much as 5.1×10^{-5} m (0.002 in); and (2) replenishable films, formed by a film transfer process from a self-lubricating composite to a metallic or possibly a ceramic bearing surface [18]. The implications of the latter idea were followed through by incorporation into powder metallurgy where products produced by powder metallurgy process were given a desired porosity, and during the process of sintering, they were impregnated with oil or desired and compatible lubricant. During contact, these lubricants are activated due to the heat generated by friction. As the temperature rises rapidly, the lubricants in the pores expand and start to move to the surface and form a thin film almost instantaneously, which prevents further metallic contact. This composite powder metallurgy process was first applied in manufacturing of bearings and was widely accepted worldwide, scientifically and commercially. This chapter discusses various aspects of self-lubricating materials from the perspective of research and engineering applicability, detailing the advancements in self-lubricating technology and analysis techniques that led to these advancements and their widespread use.

1.2   WHY SELF-LUBRICATING MATERIALS?

In most tribological applications, liquid or grease lubricants are utilized to struggle between friction and wear. The role of lubricant is to facilitate the relative motion of solid bodies by minimizing friction and wear between interacting surfaces. A lubricant made of a lower shear strength layer between the two contacting surfaces, and the shear strength of this layer is less than the surface shear strength between the sliding surfaces. Therefore, this lower shear strength lubricant layer reduces friction between the surfaces [1]. Lubricants completely can separate the surfaces without any contact between two surfaces, so no asperity junctions are formed at all. In other cases, depending on the thickness of lubricant and test condition, asperity may have contact, and it is not possible to completely avoid the asperity contact, although lubricants are able to reduce it and may also reduce the shear strength of the junctions formed [2,19,20].

The challenges for liquid lubricants arise in extreme environmental conditions such as very high or low temperatures, vacuum, radiation, and extreme contact pressure. At these conditions, solid lubricants may be the only choice and can help to decrease friction and wear without liquid lubricants. Some of the key preferences of solid lubricants in tribological applications over liquid and grease lubricants are summarized in Table 1.1 [21]. Generally, when solid lubricants are present at the contact interface, they function in the same way as their liquid counterparts. They made a low shear strength layer that can shear easily between two surfaces and avoid direct contact between surfaces. Consequently, solid lubricants can provide low friction and diminish wear damage between the sliding surfaces. In addition, a mixture of solid and liquid lubrication is likewise achievable to have a beneficial synergistic effect to enhance the friction and wear performance. Solid lubricants can be dispersed in water, oils, and greases to attain improved friction and wear properties [22,23].

Several well-known inorganic materials have lubrication properties in nature, and they are able to provide excellent tribological properties. These solid lubricants include molybdenum disulfide (MoS_2), carbonous allotropes, hBN, and H_3BO_3 [22,24,25]. The important feature of solid lubricants is a lamellar or layered crystal structure that can provide adequate lubricity. Graphite, hBN, and H_3BO_3 demonstrate the layered crystal structures.

The challenge with solid lubricants is to maintain a continuous supplier of solid lubricants on the contact surface to act as a lubricant between two sliding surfaces. Such a continuous supply of solid lubricant is more easily maintained in the case of fluid lubricants compared to solid lubricants. The most innovative development to ensure a supply of solid lubricant is to introduce the solid lubricant as reinforcement into the matrix of one of the sliding components. A self-lubricating material is one whose composition or structure facilitates low coefficients of friction and wear through the use of a self-dispensed and self-regulated lubricant delivery system such as graphite and MoS_2. These materials are becoming more and more popular as the world becomes increasingly conscious of our environmental impact and energy usage. Self-lubricating materials have the potential to effectively increase our energy efficiency through more smoothly operating system components.

Metals, ceramics, and polymers are all used as matrix materials to synthesize self-lubricating composites. Self-lubricating metal matrix composites (SLMMCs) can be processed by casting or by powder metallurgy. Almost all metals and alloys are being researched. Self-lubricating composites have been used for a long time and are utilized rather widely by the industry to combat friction and wear in a variety of sliding, rolling, and rotating bearing applications. Recent studies exhibit that some of wear particles produced at the interface are solid lubricants, and they can form a thin-film layer of the solid lubricant on the contact surfaces of materials. It causes to decrease the friction coefficient and wear rate, and enhance tribological properties. Therefore, composites reinforced by solid lubricant become self-lubricating due to the lubricant film, which prevents direct contact between the mating surfaces. Self-lubricating composites eliminate the usage of any types of external lubricants by reducing friction and wear. This lubricant film initially is not present, and it forms as a result of surface wear and subsurface deformation. They are continuously replenished by embedded solid lubricant particles in the matrix [26,27]. For example,

TABLE 1.1
Comparison of Solid and Liquid Lubricants in Tribological Applications

Application Environment and/or Condition	Solid Lubricants	Liquid and Grease Lubricants
Vacuum	Some solids (i.e., transition metal dichalcogenides) lubricate extremely well in high vacuum and have very low vapor pressure	Most liquids evaporate, but perfluoropolyalkylethers and polyalfaolefins have good durability
Pressure	Can endure extreme pressure	May not support extreme pressures without additives
Temperature	Relatively insensitive; can function at very low and high temperatures; low heat generation due to shear	May solidify at low temperatures and decompose or oxidize at high temperatures; heat generation varies with viscosity
Electrical conductivity	Some provide excellent electrical conductivity	Mostly insulating
Radiation	Relatively insensitive to nuclear radiation	May degrade or decompose over time
Wear	Provide excellent wear performance or durability at slow speeds and under fretting conditions; lifetime is determined by lubricant film thickness and wear rate	Provide marginal performance and durability at slow speeds and under fretting conditions; need additives for boundary lubrication
Friction	Extremely low friction; coefficients are feasible	Depend on viscosity, boundary films, and temperature
Thermal conductivity and heat dissipation capability	Excellent for metallic lubricants; poor for most inorganic or layered solids	Good
Storage	Can be stored for very long times (dichalcogenides are sensitive to humidity and oxygen)	May evaporate, drain, creep, or migrate during storage
Hygiene	Better industrial hygiene due to little or no hazardous emissions; because they are in solid state, there is no danger of spillage that can contaminate environment	May release hazardous emissions; liquid lubricants may spill or drip and contaminate environment; fire hazard with certain oils and greases
Compatibility with tribological surfaces	Compatible with hard-to-lubricate surfaces (i.e., Al, Ti, stainless steels, and ceramics)	Not suitable for use on nonferrous or ceramic surfaces
Resistance to aqueous and chemically aggressive environments	Relatively insensitive to aqueous environments, chemical solvents, fuels, and certain acids and bases	May be affected or altered by acidic and other aqueous environments

Source: Spalvins, T., *Inter. Confer. Metal Coat.*, Elsevier, San Diego, CA, p. 17, 1980.

aluminum/graphite composites show an improvement in lubricity, durability, and resistance to seizure under both dry and lubricated conditions [27].

In addition to metal matrix composites, a series of self-lubricating polymer and ceramic matrix composites have also been developed, tested, and offered for indus-trial use in recent years [28–30]. These composites are emerging as an important class of tribological materials, offering new means to combat friction, wear, and galling under extreme conditions.

1.3   ADVANCEMENTS IN SELF-LUBRICATING TECHNOLOGY

Friction and wear properties of a metallic body can be improved by coating a thin film of solid lubricant, which is relatively harder and more resilient onto the metallic surface. The characteristics during contact of this surface with another surface are then defined by the film layer, its hardness, homogeneity, lubricat-ing, and adhesion properties. Certain criteria that are to be met by a solid lubri-cant film coating or an embedded lubricant in a self-lubricating material are as follows [31]:

- The shear strength of the adhesion between the lubricant film and the base material must be good enough to maintain a required boundary lubrication.
- The internal cohesion of the film must be sufficiently large enough that the film does not disintegrate when subjected to friction.
- The adhesion between the particles and the layers in the shearing direction should be as small as possible to keep the resistance to friction low.

These criteria are the challenges to any coating and can only be met by selecting few self-lubricating materials such as MoS_2, hBN, H_3BO_3, graphite, PTFE, and other such solid lubricants. These solid lubricants satisfy the requirements better than the other lubricants mainly due to their varying bonding types, crystalline structures, type of base material, and process of coating or imbibing these lubricants into the material.

The initial research concerning solid lubricants led to many ideas of manufactur-ing self-lubricating materials. Some of the earliest patents and research works that were commercially applied in industries to use solid lubricant coating in manufac-turing self-lubricating materials are listed as follows:

- *1963*: The preparation and experimentation on composites composed of self-lubricating materials for varying applications began in the late 1950s and gained momentum around the mid-1960s. One such application was the development of self-lubricating composites for vacuum service. The research targeted to investigate a method to combining five variables: solid lubricant ($MoSe_2$, WSe_2, $NbSe_2$, etc.), binders (copper, silver, etc.), film former (PTFE or other suitable resin), film former/lubricant ratio, and binder combination. The research experimented with various combinations of these five vari-ables to be able to achieve the best friction and wear characteristics under sliding condition and function satisfactorily in a vacuum of 10^{-6} to 10^{-9} torr and over a temperature range from cryogenic to 400°F. This was one of the very first elaborate investigations of applying powder metallurgy to develop a composite material capable of functioning as a load-bearing surface with no lubrication other than that contained within itself. The products derived from this investigation were to be utilized as self-lubricating components for ball bearings and gears operating under a high load–low speed condition [32]. There have been many such investigations that were targeted at specific applications (mostly various types of bearings) that have tried to apply powder metallurgy to develop new self-lubricating composites [33,34].

- *1973*: Coating ferrous and aluminum metals first with nickel to a thickness of 0.0001–0.005 in. using electroless nickel chemical plating process followed by PTFE coating to a thickness of 0.0001–0.2 in. This is followed by heat treating at 320°F–820°F for 20 minutes to 2 hours to fuse the base material with nickel and PTFE, and obtain an abrasion-resistant self-lubricating surface. This process when tested proved to have improved friction and wear-resistant properties by nearly 40% as applicable to bearings [35]. This technique was followed by researches that experimented with plasma spray coatings and other such coating techniques to deposit self-lubricating films on the metal surface effectively [36].

- *1982*: Coating substrate metal by depositing a metal chalcogenide (sulfides, selenides, and tellurides of Mo, W, Nb, V, Zr, Ti, and Ta) by a process called cathode sputtering. By this time, these chalcogenides had already proved to have good lubricating properties where their crystalline structure and lamellar structure enabled these solid lubricants to provide low COF on the applied surface. In the cathode sputtering process, a base material to be coated and the corresponding chalcogenide or a mixture of chalcogenides are selected. The target material surface is bombarded with the chalcogenide ions formed in a cathode discharge tube in a rare gas environment. These ions are controlled using a magnetic field and deposited on the material surface, resulting in turbostatic deposition. A schematic of the process is shown in Figure 1.2. This process overcomes the limitations of

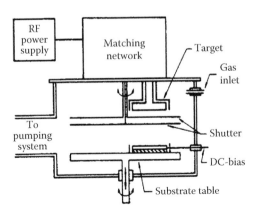

FIGURE 1.2 Schematic diagram of radiofrequency diode sputtering apparatus with a DC-biased specimen (material to be coated). (From Andersson, K.Å.B., Karlsson, S.E., Ohmae, N., *Vacuum*, 27, 379–382, 1977.)

previously available anti-friction coating techniques whereby limited wear resistance and anti-friction properties and susceptibility of the coatings to moisture are overcome. Further, some metals such as copper and bronze, which could not have been coated with solid lubricants using earlier techniques could be coated using the cathode sputtering process [37].

- *1984*: Coating fill tube with self-lubricating material that is used in medical equipment's retention valves that aid in inflation and deflation of elastic devices. The self-lubrication would then be able to tackle the problem of resistive forces encountered during the introduction of a fill tube to the device for delivery of the expansion fluid. The fill tube coating process involves cleaning the surface of the fill tube and exposing it to a fluorocarbon under ultrasonic sound field. This is followed by irradiation of the fill tube with a gamma radiation of at least 0.5 Mrad dosage. Then the fill tube is immersed in a solution comprising ethylenically unsaturated monomer and oxidizable metal ion at a temperature of ~75°C for up to 45 minutes under nitrogen atmosphere. Finally, rinsing the fill tube with deionized water leaves behind a hydrophilic lubricious coating of polymerized monomer on the surface of the fill tube [39]. This process although laborious to follow through was effective enough to be adopted for coating many other medical devices.

- *1988–1996*: Fueled by the observable applicability of self-lubricating composite materials, more researches targeted specific applications leading to increased number of patents on self-lubricating materials being filed during this period. Some of these applications were particularly directed to self-lubricating, very wear-resistant composite material for use over a wide temperature spectrum from cryogenic temperature to ~900°C in a chemically reactive environment. The applicability involved the development of more fuel-efficient engines, such as the adiabatic diesel and advanced turbo machinery. Other examples are the advanced Sterling engine and numerous aerospace mechanisms. Even though for most part the five basic variables namely, material; lubricant; bearing/part size; working pressure and sliding velocity were used to formulate self-lubricating materials, the new ideas led to the development of better versions of self-lubricating materials we know today. These consisted of combinations that included carbides, fluorides, silver, nickel-based super alloys, cobalt-based super alloys, polymer composite members, and solid lubricants selected from the group consisting of graphite, MoS_2, BN, LiF, CaF_2, NaF, and WS_2 [40–44]. This investigation done in this was also able to combine the best powder metallurgy and coating techniques to achieve a cohesive self-lubricating material that could withstand extreme working conditions.

- *Post-twentieth century*: Though most of the self-lubricating materials could be formulated using powder metallurgy techniques, it was not always feasible during mass production. During this time, coating technology gained more momentum and many developments in coating technology were made. The versatility of most vacuum deposition techniques such as plasma vapor deposition (PVD), chemical vapor deposition (CVD), and ion beam-assisted deposition (IBAD) allowed the production of complex compositions consisting of multicomponent phases [6]. The idea behind this was to still keep the mechanical properties of the base material composites and selectively coat the metal surface to improve wear resistance under severe conditions rather than just decrease friction. This concept was in a sense borrowed from tooling technology that had already adopted the PVS, CVD, and IBAD deposition techniques to coat machining tools with high wear-resistant material compositions such as (Ti, Al, V) C, N, which were optimized for this specific application [45,46]. Further research confirmed that the nature and amount of alloying elements strongly influence the friction and wear behavior of materials, and the investigations were extended to Zr, Hf, V, Nb, Cr, Mo, W, Al, and Si with various chemical combinations which are known to be self-lubricating materials. The researches conducted on amorphous carbon nitride (CN_x) coatings in the late 1990s started to gain more importance not just due to the wide applicability of wear-resistant CN_x compositions but also due to the fact that now they could be effectively deposited using the developments in coating technology [47]. Figure 1.3 shows the relation between the coating type, the coating thickness, and the coating life cycles where C and CN_x are coated on rigid magnetic disk using IBAD technique.

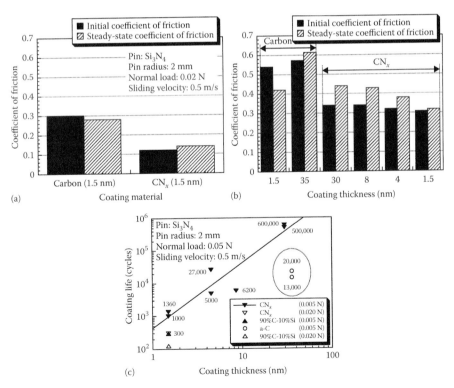

FIGURE 1.3 (a) Friction coefficient dependence on the coating type at a normal load of 0.02 N. (b) Friction coefficient dependence on the coating thickness at a normal of 0.005 N. (c) Coating life dependence on the coating thickness for C-coated and CN_x-coated disks.

TABLE 1.2
Variants of PVD and CVD Techniques

PVD Variants	CVD Variants
Cathodic arc deposition	Atmospheric pressure CVD
Electron beam physical vapor deposition	Low-pressure CVD
Pulsed laser deposition	Ultrahigh vacuum CVD
Sputter deposition	Aerosol-assisted CVD
Sublimation sandwich method	Direct liquid injection CVD
	Hot wall CVD
	Cold wall CVD
	Microwave plasma-assisted CVD
	Plasma-enhanced CVD
	Remote plasma-enhanced CVD
	Atomic layer CVD
	Combustion CVD
	Hot filament CVD
	Hybrid physical–CVD
	Metalorganic CVD
	Rapid thermal CVD
	Vapor-phase epitaxy
	Photo-initiated CVD

Though there have been advancements in PVD and CVD technologies, the basic principle remains the same in all the various techniques. The various PVD and CVD techniques developed over the past two decades are as detailed in Table 1.2.

1.4 APPLICATIONS OF SELF-LUBRICATING MATERIALS

Self-lubricating materials have proven to be very diverse in compositions that have brought about their applicability in high wear, high temperature, and high load bearing environments. However, the compatibility of formulating self-lubricating composite materials should not be considered for granted. For example, consider glass fibers with MoS_2 (fiber/MoS_2) and glass with PTFE (glass/PTFE) where whiskers or chopped fibers are dispersed in polymer matrix providing a special property. Here it would be fair to assume that the presence of MoS_2 in this specific glass fiber makes it more wear resistant. However, in fact this does not happen: the wear of fiber/MoS_2 is far greater than that of composite glass/PTFE by nearly a factor of 2 [18]. Hence, one has to conduct an in-depth analysis of proper applicability of self-lubricating materials before they are implemented. The origination of self-lubricating materials was with the applicability to various kinds of bearing materials, but soon after around the early 1990s, these materials were being tested for applicability in extreme environment conditions and even space applications [48].

Polymer composites were investigated for applicability in space tribology; even though most of the self-lubricating polymer composite gears are meant for light loads, they are found to perform well in precision space applications. Another major need for self-lubricating composites in space applications was for cryogenic bearings. Space missions require instruments that must be cooled to cryogenic temperatures. These include infrared detectors, superconducting devices, and several telescopes (infrared, X-ray, gamma ray, and high energy) [49]. In addition, high-speed turbopumps, like those used on the Space Shuttle main engines, operate with the cryogen passing directly through the bearings. Oils and greases solidify at these temperatures; therefore, the only viable alternative is solid lubrication. The NASA has been interested in developing self-lubricating composite materials for cryogenic applications since the late 1950s/the early 1960s [50,51]. In addition, the NASA has spent considerable effort while developing the bearing technology necessary for utilizing these self-lubricating composite materials in a cryogenic environment [52–56]. The testing cryogenic chamber setup adopted by NASA is shown in Figure 1.4. The best lubricant found in these early studies for use at cryogenic temperatures (and it is still the predominantly used material today) was PTFE. It was found however that PTFE had poor strength properties and tended to cold flow even under the lightest loads. It also had poor thermal conductivity, which is a problem for high-speed bearings, where heat generation can be detrimental to successful bearing operation. Thus, it was required to compound it with other materials to give it more desirable properties [48].

Today, there are hundreds of self-lubricating composite materials that are being investigated for specific applications targeting compatible mechanical and tribological properties. High-performance ceramic composites are being investigated to eliminate the earlier existing conflicts between mechanical and tribological properties such as poor mechanical properties and high friction coefficient. The new treatment techniques adopted have been able to solve this problem by using design principles of bionics and graded composites [57]. Another emerging field of prospective wide applicability from the perspective ongoing research is SLMMCs. In SLMMCs, solid lubricant materials, including carbonous materials, MoS_2, and hBN, are embedded into the metal matrices as reinforcements to manufacture a novel material with attractive self-lubricating properties. Due to their lubricious nature, these solid lubricant materials have attracted researchers to synthesize lightweight SLMMCs with superior tribological properties [58]. NiAl self-lubricating matrix composite with graphene (NSMG) material is also a new area of investigation for self-lubricating composites which have been found to be compatible with NiAl self-lubricating matrix composite providing promising surface tribological properties. Attempts are also being made to develop laser-based self-lubricating composites that coupled with additives will be able to provide a better interfacial boundary

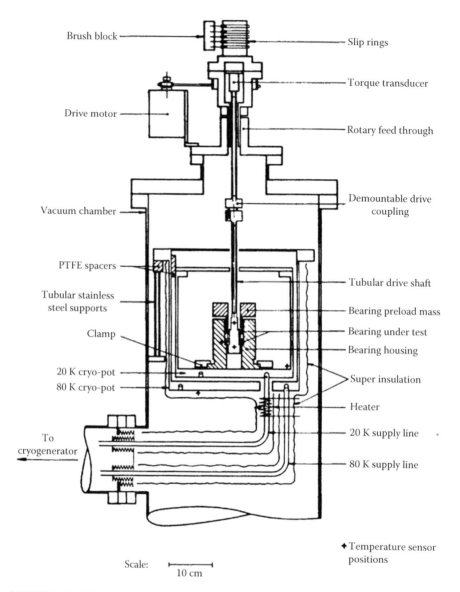

FIGURE 1.4 Schematics of cryogenic vacuum chamber and ball bearing test apparatus used by NASA for investigating self-lubricating materials. (From Khurshudov, A.G., Kato, K., *Surf. Coat. Technol.*, 86, 664–671, 1996.)

lubrication for high-pressure applications. This laser-based technique uses laser surface claddings process to apply resilient lamellar, fluorides, and oxide-based solid lubricant coatings on the metallic surfaces. The applications are listed for material coatings with major content as follows [59]:

- *Graphite*: aeronautics and astronautics, bulk metallic glasses, and aircraft turbine compressor blade tip sealant coating
- MoS_2: aeronautics, marine, automobile products, vacuum industries, food industries starved lubrication and machine tools, and plunger and pumps for oil fields
- WS_2: aerospace high-temperature applications, high-temperature adiabatic engine bearings, cylinder liners, nuclear valves, steam turbine blades, industrial gas turbines, and power generation industries
- *hBN*: high-temperature aggressive environments and metal working
- CaF_2: aerospace high-temperature self-lubrication composite coatings, compressor blades and exhaust nozzle, high-temperature wear resistance, continuous casting molds, and electrical contacts

It can be observed that the applicability of self-lubricating materials is only limited by the extent of diversity that can be achieved in the compatible composition of self-lubricating materials for a specific application. Even though the concept of self-lubricating materials has been available for several decades now, it still has not lost its potential to satisfy and improve on many engineering applications.

1.5 TECHNIQUES TO ANALYZE THE PERFORMANCE OF SELF-LUBRICATING MATERIALS

Analysis of self-lubricating materials is a characteristic of the application and type of materials being analyzed. Some integral tests that are usually performed on bearings are as follows:

- Test for the distribution of solid lubricant
- Contact pressure and distribution of edge effect
- Analysis of the behavior of wear particles and friction

The interpretation of tribological test results for a self-lubricating drive piston is shown in Figure 1.5. These data are compared with the regular components used for a drive piston to determine the effectiveness of the applied self-lubricating material.

Although mechanical testing goes hand in hand with composites of self-lubricating materials, when it comes to coated self-lubricating materials, it has to

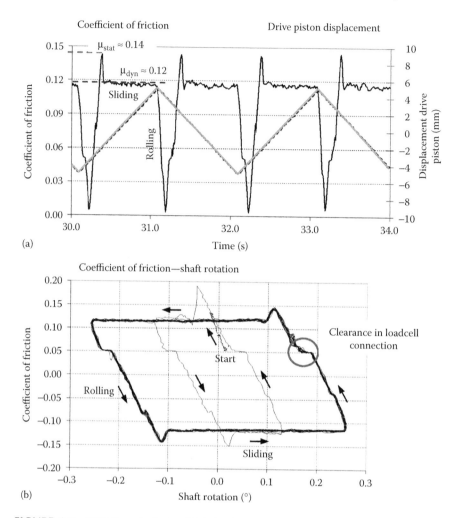

FIGURE 1.5 COF for a typical self-lubricating drive piston of an engine with respect to (a) time and piston displacement and (b) rotation of the crank shaft.

be understood that the material surface characteristics are defined by the coating methods and their effectiveness. Hence, tests that are used to analyze coatings are also adopted to test the coated self-lubricating materials. These tests include nine basic adhesion tests: pressure-sensitive tape test, acceleration (body force) testing, electromagnetic stressing, shock wave testing, tensile and shear testing, laser techniques, acoustic imaging, indentation tests, and scratch testing. Apart from the tribological tests, these tests can be further performed to assess the friction and wear behavior of the coatings. Some of the testing methods listed previously are explained in Table 1.3.

TABLE 1.3
Basic Adhesion and Film-Coating Experimental Techniques

Testing Method

Pressure sensitive tape test

Process

Single coated tapes, peel adhesion at 180° angle—A strip of tape is applied to a standard test panel (or other surface of interest) with controlled pressure. The tape is peeled from the panel at 180° angle at a specified rate, during which time the force required to effect peel is measured.

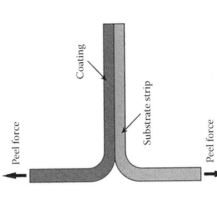

Peel force

Coating

Substrate strip

Peel force

(Continued)

TABLE 1.3 (*Continued*)
Basic Adhesion and Film-Coating Experimental Techniques

Testing Method	Process
Laser spallation techniques	The laser spallation is an experimental technique first developed to evaluate the adhesion of thin films to substrates. In this technique, an energy-absorbing coating layer is deposited onto the substrate such that it is on the opposite side of the substrate–coating interface of interest. A pulse laser, usually Nd:YAG, is made to focus on this energy-absorbing layer as seen in the figure below. This results in a sudden expansion of the layer, generating a compressive shockwave toward the substrate–coating interface. This compressive pulse strikes the interface and is partially transmitted to the coating. The reflection of this compressive pulse results in a tension pulse at the free surface of the coating, which leads to delamination when a critical pulse amplitude is applied [60].

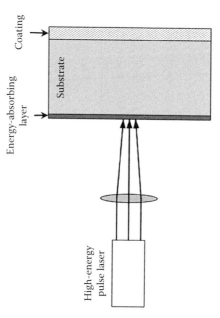

(Continued)

TABLE 1.3 *(Continued)*
Basic Adhesion and Film-Coating Experimental Techniques

Testing Method	Process
Indentation tests	When it comes to the coating/substrate system, the situation becomes more complicated because the deformation and fracture of a coated system depend not only on the coating microstructure and toughness, but also on the properties of the substrate and the coating–substrate interface [61–64]. This significantly increases the difficulty in performing the mechanical analysis and determining the corresponding mechanical properties of coatings. Despite the relatively complicated deformation and fracture behaviors occurring during the indentation test, this technique has been successfully used for adhesion evaluation of brittle coatings to ductile substrates [63–65]. The following figure shows the schematic of indentation test at the surface of a coating/substrate system and the two stages of brittle coating deformation and fracture [60].

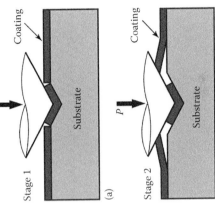

(Continued)

TABLE 1.3 (Continued)
Basic Adhesion and Film-Coating Experimental Techniques

Testing Method	Process
Scratch testing	This test is performed by drawing a spherically tipped diamond indenter across the coating surface. Meanwhile, the applied normal load on the indenter is increased until the detachment of the coating. The schematic drawing of this test is shown in the following figure [60].

(Continued)

TABLE 1.3 (Continued)
Basic Adhesion and Film-Coating Experimental Techniques

Testing Method	Process
Pin-on-disk wear test	Pin-on-disk wear testing is a method of characterizing the COF, frictional force, and the rate of wear between two materials. Multiple configurations are available depending on the goals and objectives. Common specifications include ASTM G99, ASTM G133, and ASTM F732. The experimental setup is seen in for a typical pi-on-disk wear test is as shown below. 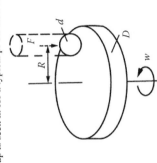

Note: F is the normal force on the pin, d is the pin or ball diameter, D is the disk diameter, R is the wear track radius, and w is the rotation velocity of the disk. There are many variations of the same experiments that perform the same functionality listed here.

REFERENCES

1. Atlas RM, Bartha R. Degradation and mineralization of petroleum in sea water: Limitation by nitrogen and phosphorous. *Biotechnology and Bioengineering.* 1972;14:309–318.
2. Atlas RM. Microbial degradation of petroleum hydrocarbons: An environmental perspective. *Microbiological Reviews.* 1981;45:180–209.
3. Momper JA. Oil migration limitations suggested by geological and geochemical considerations. *Physical and Chemical Constraints on Petroleum Migration.* 1978;A034:T1–T60.
4. Fox MF. Maintenance. In: Totten GE, (Ed). *Handbook of Lubrication and Tribology,* 2nd ed., Vol II: Theory and Design. CRC Press: Boca Raton, FL; 2006. Section: 29.
5. Holmberg K, Matthews A. Coatings tribology. In: Holmberg K, Matthews A, (Ed). *Coatings Tribology: Properties, Mechanisms, Techniques and Applications in Surface Engineering.* Elsevier Science: Amsterdam, the Netherlands; 2009;56. pp. 1–6.
6. Donnet C, Erdemir A. Historical developments and new trends in tribological and solid lubricant coatings. *Surface and Coatings Technology.* 2004;180–181:76–84.
7. Lancaster JK, Moorhouse P. Etched-pocket, dry-bearing materials. *Tribology International.* 1985;18:139–148.
8. Spalvins T. Coatings for wear and lubrication. *Thin Solid Films.* 1978;53:285–300.
9. Spalvins T. Tribological properties of sputtered MoS sub 2 films in relation to film morphology. *International Conference on Metal Coatings.* Elsevier: San Diego, CA; 1980. p. 17.
10. Ajayi OO, Erdemir A, Fenske GR, Erck RA, Hsieh JH, Nichols FA. Effect of metallic-coating properties on the tribology of coated and oil-lubricated ceramics. *Tribology Transactions.* 1994;37:656–660.
11. Bull SJ, Chalker PR. Lubricated sliding wear of physically vapour deposited titanium nitride. *Surface and Coatings Technology.* 1992;50:117–126.
12. Kimura Y, Wakabayashi T, Okada K, Wada T, Nishikawa H. Boron nitride as a lubricant additive. *Wear.* 1999;232:199–206.
13. Erdemir A. Lubrication from mixture of boric acid with oils and greases. U.S. Patent No. 5,431,830. 1995.
14. Parucker ML, Klein AN, Binder C, Ristow Junior W, Binder R. Development of self-lubricating composite materials of nickel with molybdenum disulfide, graphite and hexagonal boron nitride processed by powder metallurgy: Preliminary study. *Materials Research.* 2014;17:180–215.
15. Koskilinna JO, Linnolahti M, Pakkanen TA. Friction and a tribochemical reaction between ice and hexagonal boron nitride: A theoretical study. *Tribology Letters.* 2008;29:163–167.
16. Eichler J, Lesniak C. Boron nitride (BN) and BN composites for high-temperature applications. *Journal of the European Ceramic Society.* 2008;28:1105–1109.
17. Uğurlu T, Turkoğlu M. Hexagonal boron nitride as a tablet lubricant and a comparison with conventional lubricants. *International Journal of Pharmaceutics.* 2008;353:45–51.
18. Gardos MN. Self-lubricating composites for extreme environment applications. *Tribology International.* 1982;15:273–283.
19. Ludema KC. *Friction, Wear, Lubrication: A Textbook in Tribology.* CRC Press: Boca Raton, FL; 1996.
20. Stachowiak GW, Batchelor AW. *Engineering Tribology.* Butterworth-Heinemann: Oxford, UK; 2005.
21. Erdemir A. Solid lubricants and self-lubricating films. *Modern Tribology Handbook.* 2001;2:787.
22. Moghadam AD, Schultz BF, Ferguson J, Omrani E, Rohatgi PK, Gupta N. Functional metal matrix composites: Self-lubricating, self-healing, and nanocomposites-an outlook. *JOM.* 2014;66:1–10.

23. Rohatgi PK, Afsaneh DM, Schultz BF, Ferguson J. Synthesis and properties of metal matrix nanocomposites (MMNCS), syntactic foams, self lubricating and self-healing metals. *PRICM: Pacific Rim International Congress on Advanced Materials and Processing.* 2013;8:1515–1524.

24. Rohatgi PK, Tabandeh-Khorshid M, Omrani E, Lovell MR, Menezes PL. Tribology of Metal Matrix Composites. In: Menezes P, Nosonovsky M, Sudeep PI, Satish VK, Michael RL, (Eds.). *Tribology for Scientists and Engineers.* Springer: New York; 2013. pp. 233–268.

25. Reeves CJ, Menezes PL, Lovell MR, Jen T-C. Tribology of Solid Lubricants. In: Menezes P, Nosonovsky M, Sudeep PI, Satish VK, Michael RL, (Eds.). *Tribology for Scientists and Engineers.* Springer: New York; 2013. pp. 447–494.

26. Liu Y, Lim S, Ray S, Rohatgi P. Friction and wear of aluminium-graphite composites: The smearing process of graphite during sliding. *Wear.* 1992;159:201–205.

27. Rohatgi P, Ray S, Liu Y. Tribological properties of metal matrix-graphite particle composites. *International Materials Reviews.* 1992;37:129–152.

28. Gangopadhyay A, Jahanmir S. Friction and wear characteristics of silicon nitride-graphite and alumina-graphite composites. *Tribology Transactions.* 1991;34:257–265.

29. Prasad SV, Mecklenburg KR. Self-Lubricating aluminum metal-matrix composites containing tungsten disulfide and silicon carbide. *Lubrication Engineering.* 1994;50:511–518.

30. Fredrich K, Lu Z, Hager AM. Recent advances in polymer composites' tribology. *Wear.* 1995;190:139–144.

31. Mang T, Bobzin K, Bartels T. *Industrial Tribology: Tribosystems, Friction, Wear and Surface Engineering, Lubrication.* Wiley: Weinheim, Germany; 2011. p. 429.

32. Boes DJ, Bowen PH. Friction-wear characteristics of self-lubricating composites developed for vacuum service. *ASLE Transactions.* 1963;6:192–200.

33. Giltrow JP. Series—composite materials and the designer: Article 4. *Composites.* 1973;4:55–64.

34. Tsuya Y, Shimura H, Umeda K. A study of the properties of copper and copper-tin base self-lubricating composites. *Wear.* 1972;22:143–62.

35. Perkins G. Method of forming abrasion-resistant self-lubricating coating on ferrous metals and aluminum and resulting articles. U.S. Patent No. 3,716,348. 1973.

36. Sliney HE. Wide temperature spectrum self-lubricating coatings prepared by plasma spraying. *Thin Solid Films.* 1979;64:211–217.

37. Bergmann E, Melet G. Process for depositing on substrates, by cathode sputtering, a self-lubricating coating of metal chalcogenides and the coating obtained by this process. U.S. Patent No. 4,324,803. 1982.

38. Andersson KÅB, Karlsson SE, Ohmae N. 3.5 Morphologies of rf sputter-deposited solid lubricants. *Vacuum.* 1977;27:379–382.

39. Hyans TE. Method for forming a self-lubricating fill tube. U.S. Patent No. 4,459,318. 1984.

40. Sliney HE. Carbide/fluoride/silver self-lubricating composite. U.S. Patent No. 4,728,448. 1988.

41. Vogel FL. Motion-transmitting combination comprising a castable, self-lubricating composite and methods of manufacture thereof. U.S. Patent No. 5,325,732. 1994.

42. Rao VDN. Solid lubricant and hardenable steel coating system. U.S. Patent No. 5,484,662. 1996.

43. Peters JAD. Wear- and slip resistant composite coating. U.S. Patent No. 5,702,769. 1996.

44. Blanchard CR, Page RA. Composite powder and method for forming a self-lubricating composite coating and self-lubricating components formed thereby. U.S. Patent No. 5,763,106. 1998.

45. Knotek O, Atzor M, Prengel HG. On reactively sputtered Ti-Al-V carbonitrides. *Surface and Coatings Technology.* 1988;36:265–273.

46. Jehn HA. Multicomponent and multiphase hard coatings for tribological applications. *Surface and Coatings Technology.* 2000;131:433–440.

47. Khurshudov AG, Kato K. Tribological properties of carbon nitride overcoat for thin-film magnetic rigid disks. *Surface and Coatings Technology.* 1996;86:664–671.

48. Fusaro RL. Self-lubricating polymer composites and polymer transfer film lubrication for space applications. *Tribology International.* 1990;23:105–122.

49. Gould SG, Roberts EW. The in-vacuo torque performance of dry-lubricated ball bearings at cryogenic temperatures. *The 23rd Aerospace Mechanisms Symposium.* NASA Marshall Space Flight Center: Huntsville, AL; 1989. pp. 319–333.

50. Wisander DW, Maley CE, Johnson RL. Wear and friction of filled polytetrafluoroethylene compositions in liquid nitrogen. *ASLE Transactions.* 1959;2:58–66.

51. Wisander DW, Ludwig LP, Johnson RL. *Wear and Friction of Various Polymer Laminates in Liquid Nitrogen and in Liquid Hydrogen*, NASA TN D-3706. NASA Lewis Research Center: Cleveland, OH; 1966.

52. Scibbe HW, Anderson WJ. Evaluation of ball-bearing performance in liquid hydrogen at DN values to 1.6 million. *ASLE Transactions.* 1962;5:220–32.

53. Cunningham RE, Anderson WJ. *Evaluation of 40-Millimeter-Bore Ball Bearings Operating in Liquid Oxygen at DN Values to 1.2 Million*, NASA TN D-2637. NASA Lewis Research Center: Cleveland, OH; 1965.

54. Brewe DE, Scibbe HW, Anderson WJ. *Film-Transfer Studies of Seven Ball-Bearing Retainer Materials in 60 Deg. R (33 deg. K) Hydrogen Gas at 0.8 Million DN Value*, NASA TN D-3453. NASA Lewis Research Center: Cleveland, OH; 1966.

55. Zaretsky EV, Scibbe HW, Brewe DE. *Studies of Low and High Temperature Cage Materials.* NASA Lewis Research Center: Cleveland, OH; 1968.

56. Scibbe HW. Bearings and seals for cryogenic fluids. SAE Technical Paper No 680550; 1968.

57. Zhang Y, Su Y, Fang Y, Qi Y, Hu L. High-performance self-lubricating ceramic composites with laminated-graded structure. In: Ebrahimi F, (Ed). *Advances in Functionally Graded Materials and Structures.* InTech: Rijeka, Croatia; 2016, Chapter 4, pp 61–71. DOI: 10.5772/62538.

58. Omrani E, Dorri MA, Menezes PL, Rohatgi PK. New emerging self-lubricating metal matrix composites for tribological applications. In: Davim JP, (Ed). *Ecotribology: Research Developments.* Springer International Publishing: Cham, Switzerland; 2016. pp. 63–103.

59. Quazi MM, Fazal MA, Haseeb ASMA, Yusof F, Masjuki HH, Arslan A. A review to the laser cladding of self-lubricating composite coatings. *Lasers in Manufacturing and Materials Processing.* 2016;3:67–99.

60. Zhou K, Chen Z, Hoh HJ. Characterization of coating adhesion strength. In: Zhang SS, (Ed). *Thin Films and Coatings: Toughening and Toughness Characterization.* CRC Press: Boca Raton, FL; 2015. pp. 465–528.

61. Marot G, Lesage J, Démarécaux P, Hadad M, Siegmann S, Staia MH. Interfacial indentation and shear tests to determine the adhesion of thermal spray coatings. *Surface and Coatings Technology.* 2006;201:2080–2085.

62. Drory MD, Hutchinson JW. Measurement of the adhesion of a brittle film on a ductile substrate by indentation. *Proceedings of the Royal Society of London Series A: Mathematical, Physical and Engineering Sciences.* 1996;452:2319.

63. Hongyu Q, Xiaoguang Y, Yamei W. Interfacial fracture toughness of APS bond coat/substrate under high temperature. *International Journal of Fracture.* 2009;157:71–80.

64. Chicot D, Démarécaux P, Lesage J. Apparent interface toughness of substrate and coating couples from indentation tests. *Thin Solid Films.* 1996;283:151–157.

65. Li X, Diao D, Bhushan B. Fracture mechanisms of thin amorphous carbon films in nanoindentation. *Acta Materialia.* 1997;45:4453–4461.

2 Self-Lubricating Metal Matrix Composites

2.1 INTRODUCTION

Composite materials make use of more than one material class, with one acting as the matrix and the other as a reinforcing component. Historically, composites have been used for thousands of years, with the very first examples being mud and straw composites used for housing in Egypt and Mesopotamia (circa 1500 BC) [1]. Modern engineering composites can be found nearly everywhere in our daily life, with the most common variety being polymer matrix composites (PMCs) used, for example, as bicycle frames, hockey sticks, and airframe materials. More often than not, modern composites are designed from the perspective of mechanical properties, where the matrix alone does not have adequate strength and/or stiffness for the design requirements. The reinforcing component, which in the PMC example is often a glass or ceramic material, enhances mechanical properties by a load transfer or load sharing mechanism. Alongside the development of PMCs, scientists and engineers developed metal matrix composites (MMCs). Technologies to fabricate MMCs were developed in the 1970s with increasing adoption of these materials in various applications in the following decades [2,3]. Reinforcing components in MMCs can take the form of hard materials such as ceramics or glasses, but may also be materials for enhancement of the surface properties, known as solid lubricants.

Composite materials are often employed for tribological applications, where wear resistance and control of friction are important. Self-lubricating MMCs (SLMMCs) may be in the form of bulk composites [4,5] thick coatings or claddings [6], or thin nanocomposite coatings [7,8]. Figure 2.1 depicts the scenarios of a bulk SLMMC that contains solid lubricants to modify friction and hard phases to support load and reduce wear. Common hard ceramic phases are Al_2O_3 and SiC, and common solid lubricants are graphite (Gr), molybdenum disulfide (MoS_2), hexagonal boron nitride (hBN), and many others. The latter impart *self-lubricating* (sometimes *auto-lubricating*) properties to the MMC by forming lubricating tribofilms and, in some cases, transfer films at the contact that can be regenerated during in-service wear of the component. Here, tribofilm refers to a surface-modified layer on the material being tested, and transfer film refers to a modified layer attached to the counterface (as depicted in Figure 2.1). SLMMCs are an important class of materials for future green manufacturing and engineering sustainability. Use of SLMMCs reduces the need for oil lubrication, friction, and thus the energy consumption in the systems where they are used. They can also mitigate wear and increase the lifetime of components.

Man-made self-lubricating materials primarily involve the creation of some type of composite. A composite is a coupling of two different materials designed to inherent the qualities of both materials. Self-lubricating composites take advantage of a

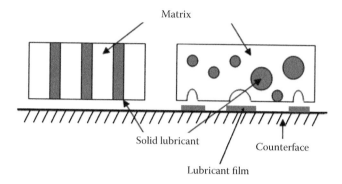

FIGURE 2.1   Schematic of self-lubricating composite and its mechanism.

hard structural matrix carefully combined with a lubricating phase. There are several ways to incorporate the lubricating phase. Dispersing solid lubricant particles or fibers throughout the matrix can be a simple and effective way to ensure that the material is constantly lubricated. The properties of the individual matrix and lubricant, the concentration of the lubricating phase, the distribution or order of the lubricating phase, and the interactions between the lubricant and the matrix are all variables that determine the quality of these types of composites. A schematic of this type of composite is shown in Figure 2.1. It shows that as the material wears against the contact surface, new solid lubricant particles will be exposed, thereby keeping the surface lubricated. A classic example of this type of composite is gray cast iron; it utilizes a hard iron matrix with dispersed lubricating graphite flakes. Constructing a composite with alternating layers of the structural phase and the lubricating phase is also an effective way to engineer self-lubricating materials.

As mentioned earlier, the unique nature of self-lubricating composites is that the wear particles formed on the contact surface act as solid lubricants, and they can reduce the COF and wear rate. For instance, under sliding conditions, the metal/graphite composite is self-lubricating because of the transfer layer of the graphite on the tribo-surfaces and then the formation of a thin layer of graphite, which prevents direct contact between the mating surfaces [9]. To have an effective lubricant layer, it is also important that the solid lubricant have a strong adhesion on the bearing surface; otherwise, this lubricant layer can be easily rubbed away and can tend to very short service life.

SLMMCs have been extensively employed when liquid lubricants are impractical. The particulate composites of solid lubricant and the matrix alloy are characterized by (1) the composition and microstructure of the matrix alloy; (2) the size, volume fraction, and distribution of particles; and (3) the nature of the interface between the matrix and the dispersed graphite [9,10].

In SLMMCs, solid lubricants apparently play a key role in enhancing the tribological properties of composites through the formation of a solid lubricant-rich film on the tribo-surface which acts as solid lubrication to improve tribological properties [11,12]. Due to low shearing strength of solid lubricant layers, solid lubricant particles entrapped immediately between two contact surfaces will shear, and it is the phenomenon that helps to form this lubrication layer. This lubricant film can assist composites to have improved tribological properties by (1) reducing the magnitude of shear stress transferred to the material underneath the contact area, (2) reducing the plastic deformation in the subsurface region, (3) preventing a direct metal-to-metal contact, and (4) finally acting as a solid lubricant between two sliding surfaces. Therefore, these strong reasons are able to help self-lubricating composites have better friction, wear, and seizure resistance than alloys. The important factors for this lubrication film are composition, area fraction, thickness, and hardness. The larger thickness of the lubrication film at the sliding surface layer can be achieved at higher amount of solid lubricants embedded into the matrix, which is responsible for playing an effective role in keeping the friction and wear behavior of the composite low. The characteristics of the lubricant film on the contact surfaces depend on the matrix properties, adhesion of the solid lubricant film to the matrix, and the presence of an environment that permits graphite to spread in the form of a film and act as a solid lubricant [9,10].

2.1.1 Effect of Solid Lubricants in Metal Matrix Composites

There is a long history of incorporation of solid lubricant materials into metal matrices. Graphite inclusions are more extensively studied, with Rohatgi and coworkers carrying out much of the early work on Al–Gr composites fabricated by casting routes [2,11–15]. Researchers have also incorporated MoS_2, WS_2, and hBN into metals by powder metallurgy or other techniques [16–18]. Since the early years of manufacture of SLMMCs, which was primarily using casting and powder metallurgy methods, there have been significant advances in both alternative processing technologies and the materials themselves (e.g., Gr replaced by carbon nanotubes [CNTs] or G). In the end, regardless of the processing route or material system, the fabrication of any SLMMC has a few key goals in mind. First, the solid lubricant should be dispersed as homogeneously as possible and be unmodified by the processing. The actual volume fraction of the solid lubricant included is often a function of the application itself. This is because as the volume fraction of the solid lubricant increases, there is inevitably a detriment to the bulk mechanical properties. Often there may be some benefit to the mechanical properties at low volume fractions, but at higher concentrations, the solid lubricant will result in softening. Thus, the second main goal is to find a balance of the lubricant content. Ideally, one requires a sufficient amount that there can be a sustainable lubricating tribofilm at the material's surface throughout its lifetime, but not so much lubricant that there is an unacceptable debit in mechanical properties. This is why one often finds SLMMCs that also include hard phases. Load-bearing nature of hard inclusions helps to overcome the debit in mechanical properties introduced by the solid lubricant. Many of the newly developed processing routes also seek to incorporate higher content of lubricant with better maintainability of mechanical properties. The remainder of this chapter will address the tribology of carbon-based and MoS_2hBN/WS_2/CaF_2/BaF_2 SLMMCs with special attention to the role of third bodies in determining the performance of these materials.

2.2　ALUMINUM MATRIX COMPOSITES

With increasing demands for high-performance materials for use in energy-efficient, low-maintenance engineering systems, lightweight, high-strength aluminum alloys are of great interest to designers [19]. Excellent properties of aluminum alloys and aluminum matrix composites (AMCs) such as high specific strength [20], high corrosion resistance [21], good thermal conductivity [22], low electrical resistivity [23], and high damping capacity [24] tend to increase their application in several industries such as aerospace, marine, and automotive for components, such as engine, cylinder blocks, pistons, and piston insert rings [25]. The best substitution for cast iron and bronze alloys is AMCs when tribology is the main concern because of their superior wear and seizure resistance [26,27].

Generally, aluminum matrix reinforced by graphite particles exhibit improvement in tribological properties in comparison with aluminum composite reinforced with other ceramic particles such as Al_2O_3 and SiC [25]. Decrease in friction and wear rate is significant in the presence of graphite as reinforcement compared to unreinforced matrix alloys as a result of the incorporation of graphite particles [28]. Ames et al. [29] investigated the effect of graphite on the wear regime of composites. The results showed that composites reinforced by graphite did not exhibit any severe wear and continued to remain in the mild wear regime even at high normal load conditions, whereas its alloy observed the severe wear regime at high normal load. The reason for this behavior is tribofilm formation during sliding that can provide adequate lubrication between contact surfaces (Figure 2.2).

Important factors can affect the tribological properties are the type [30], size [31], volume fraction [31], and distribution [32] of the reinforcing phase as well as the manufacturing technique of the composites. Another important factor that can affect

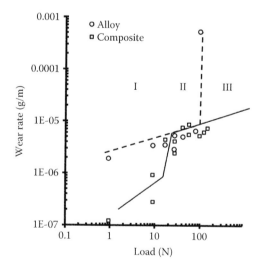

FIGURE 2.2　Effect of graphite particles on the transition point of MMCs. (From Ames, W. and Alpas, A., *Metall. Mater. Transac. A.*, 26, 85–98, 1995.)

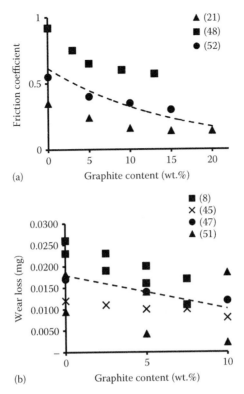

FIGURE 2.3 Correlation between graphite content and (a) friction coefficient (From Akhlaghi, F. and Pelaseyyed, S.A., *Mater. Sci. Eng. A.*, 385, 258–266, 2004; Akhlaghi, F. and Zare-Bidaki, A., *Wear*, 266, 37–45, 2009; Akhlaghi, F. and Mahdavi, S., *Ad. Mater. Res.*, 264–265, 1878–1886, 2011.) and (b) wear loss. (From Ravindran, P. et al., *Mater. Des.*, 51, 448–456, 2013; Shanmughasundaram, P. and Subramanian, R., *Ad. Mater. Sci. Eng.*, 1–8, 2013; Ravindran, P. et al., *Ceramics Inter.*, 39, 1169–1182, 2013.)

the mechanical properties and tribological behavior of aluminum/graphite composites is the interface between the matrix and graphite [33,34]. Furthermore, test parameters such as load and sliding velocity can affect the tribological properties of self-lubricating aluminum/graphite composites.

The effect of volume fraction on the coefficient of friction (COF) and wear rate is exhibited in Figure 2.3a and b, respectively. As shown in the figure, there is an acceptable effect on the tribological properties where the friction coefficient and wear rate of aluminum/graphite self-lubricating composites decrease with an increase in the amount of graphite [35–40] due to an increasing in the thickness of the lubricating film at the interfaces with increasing graphite content, which consequently decrease the contact of asperities [39,41]. This phenomenon occurs as a consequence of the shearing of graphite particles placed between the sliding surfaces of the composite and the formation of thin lubricant film. It reduces the magnitude of shear stress and plastic deformation in the subsurface region as well as avoids the direct metal-to-metal contact. This lubricant layer also acts as a solid lubricant between the two sliding surfaces [42].

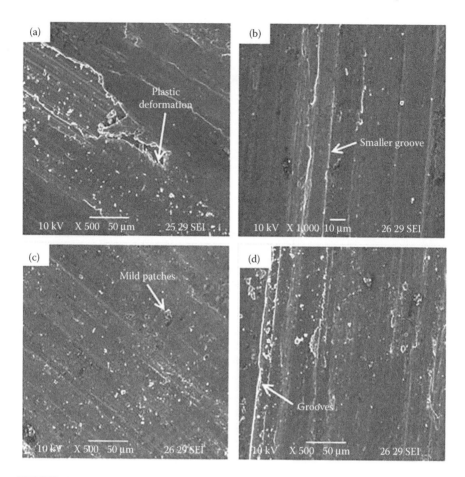

FIGURE 2.4 SEM image of the worn surface of (a) unreinforced aluminum A7075, (b) A7075/5 wt.% graphite, (c) A7075/10 wt.% graphite, and (d) A7075/20 wt.% graphite. (From Baradeswaran, A. and Perumal, A.E., *Composites: Part B*, 56, 464–471, 2014.)

Figure 2.4 depicts the worn surface of unreinforced aluminum and aluminum reinforced by graphite. It is obvious that the grooves of composites are smaller than the unreinforced alloys when comparing worn surface of unreinforced aluminum alloys (Figure 2.4a) with the composites reinforced by 5, 10, and 20 wt.% graphite (Figure 2.4b–d). Deep abrasion grooves were observed at the sliding surface, which is a result of severe plastic deformation. Additionally, the surface of composites shows a graphite layer where the thickness of tribofilm increases with increasing graphite volume fraction in the composite. This graphite tribolayer acts as a protective layer to preclude direct contact between the composite surface and the counter surface. Thus, graphite layer on the surface can effectively reduce the friction coefficient of the MMC reinforced by graphite particles [42,45].

Generally, the wear rate and friction coefficient of aluminum/graphite composites increases with applying higher load as shown in Figure 2.5a and b

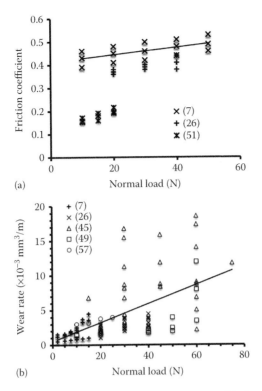

FIGURE 2.5 Correlation between the applied load and (a) the friction coefficient. (From Ravindran, P. et al., *Ceramics Inter.*, 39, 1169–1182, 2013; Srivastava, S. et al., *Inter. J. Mod. Engi. Res.*, 2, 25–42, 2012; Radhika, N. et al., *Indus. Lub. Tribol.*, 64, 359–366, 2012.) and (b) the wear rate. (From Suresha, S. and Sridhara, B.K., *Comp. Sci. Technol.*, 70, 1652–1659, 2010; Srivastava, S. et al., *Inter. J. Mod. Engi. Res.*, 2, 25–42, 2012; Basavarajappa, S. et al., *J. Mater. Engi. Perfor.*, 15, 668–674, 2006; Radhika, N. et al., *Indus. Lub. Tribol.*, 64, 359–366, 2012; Baradeswaran, A. and Elayaperumal, A., *Ad. Mater. Res.*, 287–290, 998–1002, 2011.)

[35,37–39,41,46–51]. Due to higher plastic deformation at higher normal loads, the wear rate increases, and hence delamination wear occurs [35]. Figure 2.6 is the scanning electron microscope (SEM) image of worn surface at different normal loads, which shows the nature of wear mechanisms. The dominant wear mechanism at low loads is abrasion due to the presence of grooves formed by the reinforcing particles on the worn surfaces of the composites along the sliding direction. With increasing normal loads, the grooves on the contact surface become deeper. Conversely, delamination is the dominant wear mechanism at higher normal loads due to the fracture strength of the particles less than the applied stresses [46]. Accordingly, at very high loads, the plastic flow of the material becomes dominant. Extensive plastic deformation causes severe wear, which is the key reason for severe wear conditions occurring at higher loads [48]. Generally, parallel ploughing grooves and scratches can be seen over all the surfaces in the direction of sliding [39]. Figure 2.7 shows the effect of embedding

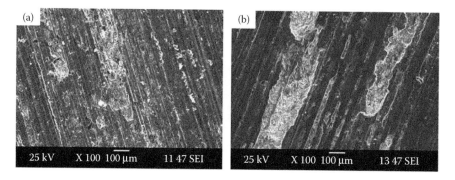

FIGURE 2.6   SEM of the surface of composite (Al/9%Al$_2$O$_3$/3%Gr) for (a) load = 20 N and V = 1.5 m/s and (b) load = 40 N and V = 1.5 m/s. (From Radhika, N. et al., *Indus. Lub. Tribol.*, 64, 359–366, 2012.)

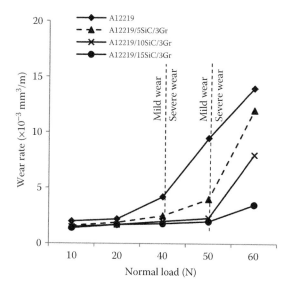

FIGURE 2.7   Variation of the wear rate with applied load at a sliding speed of 3 m/s for a sliding distance of 5000 m. (From Basavarajappa, S. et al., *J. Mater. Engi. Perfor.*, 15, 668–674, 2006.)

of graphite on the transition point. The transition point for the alloy occurs at the load of 40 N and then the wear regime enters the severe wear regime. However, the transition point for the composite containing 5SiC/3Gr occurs at the load of 50 N, whereas the severe wear regime is not present up to the normal load of 60 N for the Al/15SiC$_p$/3Gr composite [46].

It is generally accepted that when the particle size of reinforcements decreases at a constant volume fraction, it influences strength, ductility, machinability, and fracture toughness [52–55]. Even though few studies have investigated the effect of graphite particle size on tribological properties, its effect on mechanical properties has been explored extensively [35,42,45,56–59]. Jinfeng et al. [60] investigated the effect of

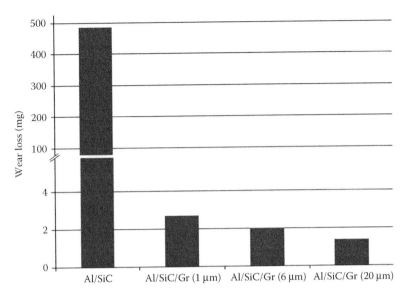

FIGURE 2.8 Wear loss of the composites for Al/SiC and Al/SiC/Gr. (From Jinfeng, L. et al., *Rare Metal Mater. Engi.*, 38, 1894–1898, 2009.)

different particle sizes (average diameter: 1, 6, and 20 μm) at a constant volume fraction on the wear rate of Al/SiC/Gr hybrid composites as shown in Figure 2.8. It has been demonstrated that the wear loss of Al/SiC/Gr composites gradually reduced with increasing graphite particle size, from 2.7 mg wear loss for a graphite particle size of 1 μm to 1.4 mg wear loss for a graphite particle size of 20 μm. Thus, Al/SiC/Gr composites with finer graphite particles exhibit lower wear resistance [60].

The greatest disadvantage of embedding aluminum graphite composites is the low mechanical properties of self-lubricating composites reinforced by graphite. By increasing the amount of graphite in the matrix, the mechanical properties of AMCs reinforced by graphite decrease [36,42,43] as shown in Figure 2.9. Many approaches are employed to reduce the damaging effect of graphite particles on the deterioration of mechanical properties. Two well-known methods are using hybrid aluminum MMCs containing ceramic particles and graphite particles, and embedding nano-sized carbounous materials such as CNTs and nano-graphite or graphene [61] in order to achieve improved mechanical, electrical, and tribological properties.

Previous studies reveal excellent tribological properties of metal/CNT composites due to the lubricating nature of CNTs. In the same manner of graphite, CNTs can form a lubricant film between the contact surfaces during sliding and reduce the direct contact. There is a weak van der Waals bond between CNT and metal, which is directed to effortlessly slide or roll between sliding surfaces and diminish the direct contact between the surfaces. Consequently, composites reinforced by CNT show a decrease in friction coefficient of the composite. The enhancement in wear resistance is due to the characteristic of CNTs as spacers between contact surfaces that avoid direct contact between asperities of rough surfaces [62]. Generally, several material parameters, such as the amount of reinforcements, the size of reinforcement, and the

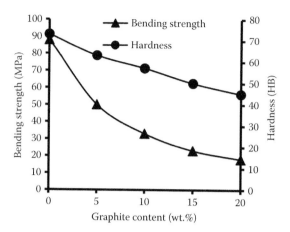

FIGURE 2.9 Effect of graphite content on the mechanical properties of MMC. (From Akhlaghi, F., Zare-Bidaki, A., *Wear*, 266, 37–45, 2009.)

spatial distribution, have direct effect on tribological properties of self-lubricating metal/CNT composites [61,63].

Zhou et al. [64] have investigated the mechanical and tribological properties of Al-Mg/multiwalled CNT (MWCNT). As shown in Figure 2.10a, addition of MWCNT into Al-Mg alloy improved the hardness of the composites compared to unreinforced aluminum alloy, and by increasing the volume fraction of MWCNTs, the hardness of the composites initially increased and then decreased at higher volume fraction of MWCNTs. The effect of the volume fraction of MWCNTs on the friction coefficient and the wear rate of the composite is shown in Figure 2.10b. The COF and wear rate decreased even at high volume fraction of MWCNTs. X-ray diffraction analysis of contact surface reveals that the wear particles are mainly aluminum oxide. During the wear process, laminated oxide films were formed at contact surfaces that they subsequently broke up and flaked off due to low adhesion between the oxide films and the aluminum matrix. The oxide particles that were present at contact surfaces are harder than the aluminum matrix and were able to increase the abrasive wear. As the aluminum matrix gradually wore out during the sliding process, the CNTs that were initially embedded in the matrix now are pulled out and exposed on the contact surface and form a lubricant film on the worn surface. These solid lubricating films significantly reduce the adhesive wear caused by oxide particles compared to unreinforced aluminum.

Choi et al. [65] have investigated the effect of the test parameter, including normal load and sliding speed, on the COF and wear rate. Figure 2.11 demonstrates the effect of the applied load and sliding velocity on COF and wear loss. Investigations have presented that the COF and wear loss increase with increasing normal load for aluminum/4.5 vol.% MWNT composite at a constant sliding speed of 0.12 m/s. Nevertheless, the COF is still lower than 0.1. At higher applied load, the friction coefficient and wear loss are increased as severe wear plays the dominant wear mechanism, and as a result, severe surface damage is observed at higher applied loads. However, the COF and wear loss slightly has decreased with increasing sliding speed at a constant applied load of 30 N.

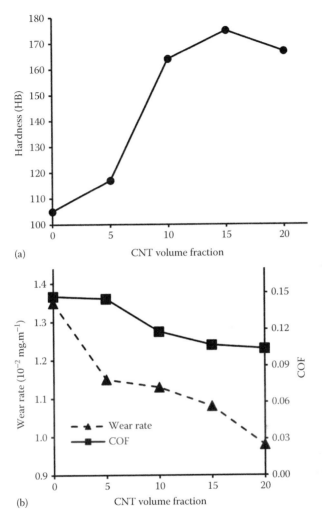

FIGURE 2.10 Variations of (a) Brinell hardness (HB) and (b) wear rate and COF with MWNT content for melt-infiltrated MWNT/Al–Mg composites under an applied load of 30 N and a sliding velocity of 1.57 m/s. (From Zhou, S.-M. et al., *Composites: Part A*, 38, 301, 2007.)

Ghazaly et al. [66] synthesized AA2124/graphene nanocomposites and investigated the effect of weight percentages (0.5, 3, and 5 wt.%) of graphene on mechanical and tribological properties. As shown in Figure 2.12, self-lubricating composites reinforced by 3 wt.% graphene exhibited the best tribological properties under dry wear test in comparison with unreinforced and other nanocomposites. Longitudinal grooves in all samples were observed on worn surfaces of unreinforced aluminum alloy and aluminum/graphene nanocomposites (Figure 2.13). Furthermore, the size of scratches, craters, and delamination of AA2124/3 wt.% graphene composite are smaller than those of unreinforced aluminum alloy and aluminum/graphene

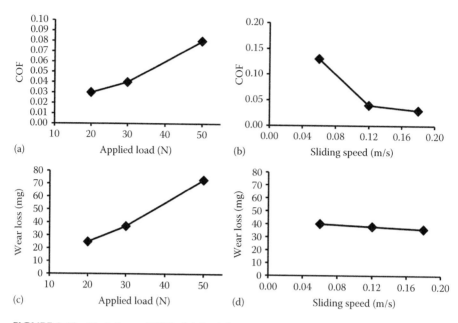

FIGURE 2.11 Variations of COF of 4.5 vol.% MWNT/Al composite with (a) applied load at a sliding speed 0.12 m/s and (b) sliding speed at an applied load of 30 N. Variations of wear loss of 4.5 vol.% MWNT/Al composite with (c) applied load at a sliding speed of 0.12 m/s and (d) sliding speed at an applied load of 30 N. (From Choi, H.J. et al., *Wear*, 270, 12, 2010.)

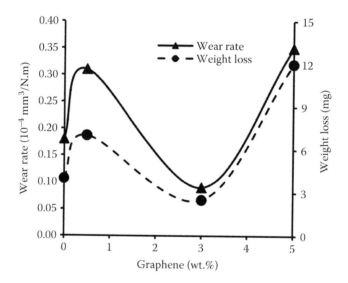

FIGURE 2.12 Wear rate and weight loss variation as a function of graphene content in AA2124 matrices. (From Ghazaly, A. et al., *Light Metal.*, 2013, 411–415, 2013.)

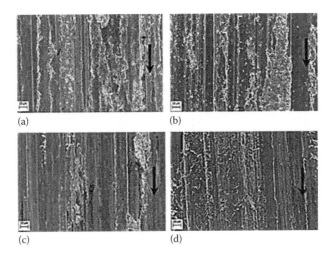

FIGURE 2.13 SEM micrographs of worn surfaces of AA2124: (a) unreinforced, (b) 0.5, (c) 3, and (d) 5 wt.% graphene nanocomposites. (From Ghazaly, A. et al., *Light Metal.*, 2013, 411–415, 2013.)

nanocomposites. The wear regime of AA2124 was severe, whereas the wear regime was changed to mild for AA2124/3 wt.% graphene nanocomposite. Shallow parallel grooves and ridges were formed on the worn surfaces of AA2124/0.5 and 5 wt.% graphene nanocomposite due to microploughing. Thus, the dominant wear mechanism is severe plastic deformation of the matrix that results in high wear rate. At higher magnification, the worn surfaces of AA2124 alloy are containing debris, whereas there is no wear debris on the worn surfaces of nanocomposite as illustrated in Figure 2.14. Alumina fragmented films or strain hardened particles are the two main sources of debris. Detached consolidated powders were formed as wear debris on contact surfaces. By comparing the worn surfaces at high magnifications, nanocomposite containing 3 wt.% graphene exhibits smoother surface between other materials. Additionally, AA2124/3 wt.% graphene composite is covered by lubricant films on the surface, and it tends to decrease the friction and wear rate due to the soft nature of the lubricant film. Therefore, shallow grooves and mild damage occur on the worn surfaces of AA2124/3 wt.% graphene composites along the sliding direction, whereas the surface of AA2124/5 wt.% graphene exhibits deep grooves and severe damage. Hence, a significant increase in wear rates and weight loss of AA2124/5 wt.% graphene occurs.

Zamzam [67] was synthesized aluminum/3graphite, aluminum/5MoS$_2$, and aluminum/3graphite/2MoS$_2$. The samples were tested in two different conditions: as-extruded and annealed. Figure 2.15 compares the weight loss of different composites reinforced by different solid lubricants and also shows the effect of annealing on weight loss. The as-extruded samples have shown better tribological properties than those tested after annealing. The reason for negative effect of annealing is that the oxidation of MoS$_2$ occurs at 573°F generating MoO$_3$, whereas the oxidation of graphite occurs at 723°F forming CO and CO$_2$. The results reveal that aluminum

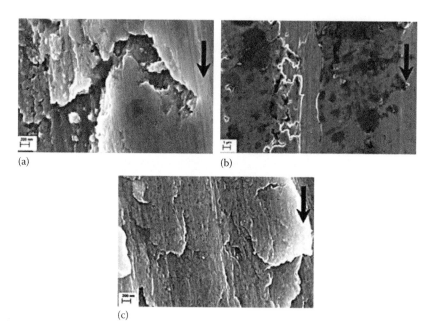

FIGURE 2.14　SEM micrographs of worn surfaces of AA2124: (a) 0, (b) 3, and (c) 5 wt.% graphene nanocomposites. (From Ghazaly, A. et al., *Light Metal.*, 2013, 411–415, 2013.)

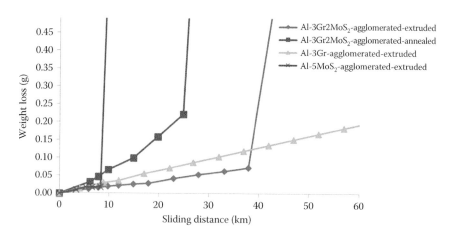

FIGURE 2.15　Effect of the types of solid lubricant on weight loss of aluminum composites. (From Zamzam, M., *Mater. Transac.*, JIM, 30, 516–522, 1989.)

reinforced by graphite possesses better wear resistance than that reinforced by MoS_2. In addition, aluminum/MoS_2 and aluminum/graphite/MoS_2 show a transition point from mild wear to severe wear at 8 and 38 km, respectively, whereas aluminum/graphite does not exhibit any failure and severe wear up to 60 km. This phenomenon can be attributed to the increase in the temperature of the contact surface and the

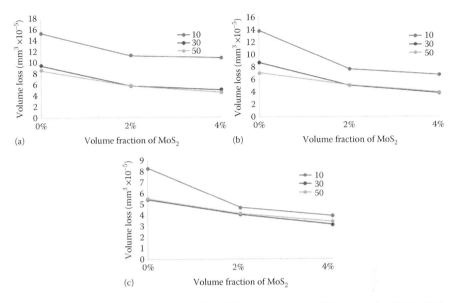

FIGURE 2.16 Wear rate of Al-Si$_{10}$Mg/Al$_2$O$_3$/MoS$_2$ composite at sliding speeds of (a) 2, (b) 3, and (c) 4 m/s. (From Dharmalingam, S. et al., *J. Mater. Engi. Perfor.*, 20, 1457–1466, 2011.)

oxidation of solid lubricants. When oxidation occurs, severe wear starts and failure of pin is noticeable. When the oxidation temperature of graphite is higher, it means more stability than MoS$_2$ and can postpone the failure.

Dharmalingam et al. [68] studied the tribological properties of Al-Si$_{10}$Mg reinforced by alumina (10–20 μm) as hard reinforcement and MoS$_2$ (1.5 μm) as soft reinforcement. The volume fraction of alumina is 5 wt.%, and that of MoS$_2$ is varying between 2 and 4 wt.%. The effects of the volume fraction of MoS$_2$, applied load, and sliding speed were investigated. It is evidently detected from Figure 2.16 that the lowest specific wear rate is observed in the composite reinforced by 4 wt.% of MoS$_2$ at 30 N applied load and 4 m/s sliding speed. From Figure 2.17, the lowest COF is obtained for the composite reinforced by 4 wt.% of MoS$_2$ at 10 N applied load and 4 m/s sliding speed. The results can be explained by the SEM micrographs of the worn surface of aluminum composite specimens. The composites reinforced by 4 wt.% MoS$_2$ clearly exhibit the fine grooves with fewer plastic deformations at 10 N applied load and 4 m/s sliding speed as shown in Figure 2.18a. In contrast, at higher applied load (50 N) and the same sliding speed (4 m/s), grooves on the worn surface of the composites reinforced by 4 wt.% MoS$_2$ become deeper and wider, and the plastic deformation at the edge of grooves is high as shown in Figure 2.18b. Many researchers stated that the oxidative wear was the dominant wear mechanism that occurs during dry sliding wear of aluminum composites. The oxide layer of lubricant phenomena is improved with the increase in the amount of MoS$_2$, suggesting that the worn-out surface is oxidized. In aluminum–alumina–MoS$_2$ composites, more oxide is formed at higher sliding speed for a constant applied load, which leads to an increase in metal and MoS$_2$, thus lowering the specific wear rate.

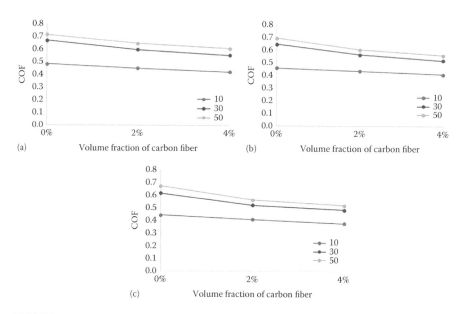

FIGURE 2.17 Wear rate of Al-Si$_{10}$Mg/Al$_2$O$_3$/MoS$_2$ at sliding speeds of (a) 2, (b) 3, and (c) 4 m/s. (From Dharmalingam, S. et al., *J. Mater. Engi. Perfor.*, 20, 1457–1466, 2011.)

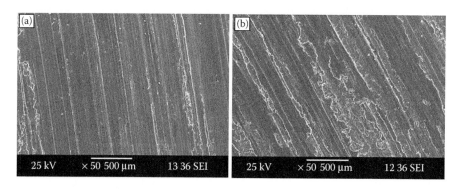

FIGURE 2.18 SEM image of the worn surface of aluminum–5 wt.% of alumina–4 wt.% of MoS$_2$ at 4 m/s: (a) 10 N and (b) 50 N. (From Dharmalingam, S. et al., *J. Mater. Engi. Perfor.*, 20, 1457–1466, 2011.)

2.3   COPPER MATRIX COMPOSITES

The advantages of copper composites reinforced by solid lubricant are excellent thermal and electrical conductivities and small thermal expansion coefficient, while they can retain properties of copper. Due to these good properties and self-lubricating performance of copper/solid lubricant composites, they have been extensively used as contact brushes and bearing materials in many industrial applications. However, in the case of low voltage and high current densities, typically for sliding parts of

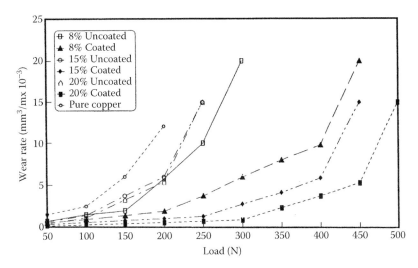

FIGURE 2.19   Wear rate vs. applied normal load of pure copper and Cu-coated and uncoated graphite composites of graphite contents of 8%, 15%, and 20%. (From Moustafa, S. et al., *Wear*, 253, 699–710, 2002.)

welding machines, it is required to employ materials with a very high specific electrical conductivity, good thermal conductivity, and low friction coefficient. Such conditions are fulfilled only by copper/graphite composite material. Copper matrices containing graphite composites are widely used as brushes and bearing materials in many applications.

Moustafa et al. [69] investigated the effect of copper content (8, 15, and 20 wt.%) at different normal loads (50–500 N) synthesized by powder metallurgy. They used either Cu-coated graphite powders or a mixture of copper and graphite powders. The pin-on-ring tribometer was employed for wear testing. Figure 2.19 shows the comparison of volumetric wear rates for copper/graphite composites with pure copper at different normal loads and coated with uncoated graphite. The copper/graphite composite shows lower wear rate than sintered copper compacts up to 200 N normal load due to the presence of a smeared graphite layer on the contact surface of the wear sample, which is produced by extrusion of graphite between the surface of the tested pin during sliding. This lubricant layer acts as a solid lubricant, and it tends to reduce the wear rate. Composites containing 8% and 15% graphite can endure the highest loads up to 450 N, whereas composites containing 20% graphite can withstand up to 500 N applied normal load. Figure 2.20 shows the variation of COF at different applied loads for pure copper, Cu-coated, and uncoated graphite self-lubricating copper composites. The influence of graphite on the COF is also significant due to the presence of the graphite layer at the sliding surface of the wear sample, which acts as a solid lubricant. In addition, the coated graphite composites possess the lowest COF than those of uncoated composites. Results reveal the reduction in the COF by an increase in the graphite content in either cases of coated and uncoated composites.

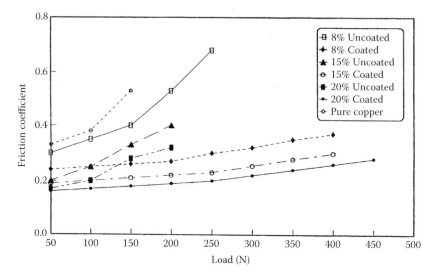

FIGURE 2.20 COF against applied load for pure copper and Cu-coated and uncoated graphite composites of graphite contents of 8%, 15%, and 20%. (From Moustafa, S. et al., *Wear*, 253, 699–710, 2002.)

A thicker and denser lubricant film was observed in higher amount of graphite. Therefore, an increase in the volume fraction of graphite content in the coated graphite composites tends to a significant reduction in the wear rate for all graphite contents. The results show that coated graphite composites become high adherent and compacted than uncoated graphite composites because of the high mismatch and good contact between graphite and copper matrix. Accordingly, weak bond between graphite and copper matrix leads to the fast removal and regeneration of a layer during sliding wear test. Hence, copper/uncoated graphite has higher wear rate and COF than copper/coated graphite at each corresponding graphite volume fraction.

Zhao et al. [70] investigated the effect of graphite reinforcement on copper matrix and tried to find the wear mechanism and the reason for this mechanism by doing SEM and energy dispersive X-ray (EDX) analysis. The results of COF and wear rate are shown in Figure 2.21. The friction coefficient and wear rate of self-lubricating copper composites decrease with increasing volume fraction of graphite. The slippage between the layers of graphite proceeds easily at a test load because of the hexagonal structure. Graphite sticks to the worn surface, and it tends to form a solid self-lubricating layer on the worn surface. It causes to greatly improve the wear properties of Cu–graphite composites in comparison with those of pure copper due to reducing direct contact between two surfaces and transforming contacts between metal and metal into the contacts between graphite film and metal or graphite film and graphite film.

Figure 2.22a and b shows the SEM and EDX analysis on the worn surface of pure copper and Cu–graphite composites, respectively. When the harder material (disk) contacts some asperities on the surface, adhesion happens at the interface of pure copper sample, whereas less adhesion takes place on the surface of copper/graphite

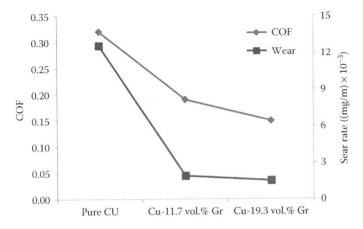

FIGURE 2.21 Microhardness and wear properties of Cu and Cu–graphite composites. (From Zhao, H. et al., *Compos. Sci. Technol.*, 67, 1210–1217, 2007.)

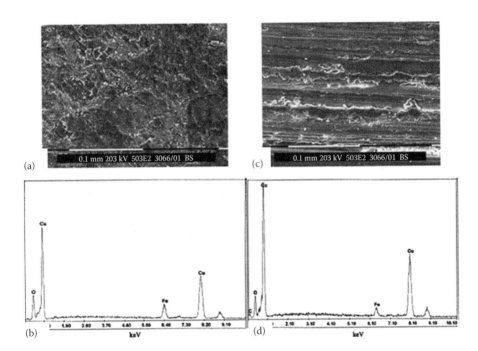

FIGURE 2.22 (a) SEM micrograph, (b) EDX analysis of the worn surface of pure copper, (c) SEM micrograph, and (d) EDX analysis of the worn surface of Cu/graphite. (From Zhao, H. et al., *Compos. Sci. Technol.*, 67, 1210–1217, 2007.)

and results in less split fragments. Adhesion hinders the easy sliding between the pure copper pin and the disk, which results in a large friction coefficient and wear mass loss. Near the adhesion areas, plastic deformation is observed, and micro-cracks and voids are formed at the surface (Figure 2.22a) [71], whereas less plastic deformation and ploughing grooves along the sliding direction are detected on the worn surface of Cu/graphite composites as shown in Figure 2.22b. The micro-cracks grow and extend to the surface of pure copper, and then many copper debris produce on the worn surface as shown in Figure 2.22a. However, the graphite film formed on the surface causes to change the contact areas between the disk and the copper matrix to the graphite film and disk. In addition, the graphite film has weak bonding between its layers, and it leads to the distribution of the lubricant film on the worn surface and the disk. Accordingly, for copper/graphite, the wear mecha-nism is changed into delaminating wear. In addition, some copper debris is present on the surface of the disk, whereas there is no obvious transfer of the composites on the disk. In addition, mild oxidative wear also occurs in terms of O content as shown in Figure 2.22c. However, the amount of Fe and O on the worn surface of the composites decreases in comparison with pure copper in terms of EDX analysis as illustrated in Figure 2.22d. Consequently, graphite particles can enhance the tribological properties and reduce the oxidative wear.

In addition, other researches show improvement in tribological properties of copper matrix by embedding graphite. Kovacik et al. [63] investigated the effect of graphite at a high volume fraction up to 50 vol.%. As graphites have good lubri-cant properties, almost embedding small volume fraction of graphite into copper matrix composite tends to significant improvement of COF of copper composites up to a critical point. It was confirmed that there is a critical point where before this point, with increasing concentration of graphite, the COF and wear rate of compos-ites decrease, whereas after this point, the COF of composites becomes independent of the volume fraction of graphite (Figure 2.23). For MMCs, the critical point is

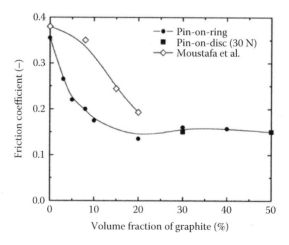

FIGURE 2.23 Composition dependence on the friction coefficient of uncoated Cu–graphite composites and coated (right) at 100 N. (From Kovacik, J. et al., *Wear*, 256, 417–421, 2008.)

not about 20 vol.% of graphite [72]. However, it significantly depends on the matrix and reinforcement. For example, it is 12 vol.% of uncoated graphite for fine graphite powder (16 µm), whereas it is 23 vol.% of uncoated graphite for coarse powder (25–40 µm, [69]). For coated graphite, the critical point was reported above 25 vol.% of graphite. The reason is that at low graphite volume fraction, discontinuous graphite film is formed on the surface, whereas at increasing volume fraction of graphite, the graphite film becomes less separated, and by increasing the amount of graphite, there is a homogeneous graphite film that covers the whole area of pin. This tends to form a graphite-rich mechanically mixed layer (MML) between the contacting surfaces. A large amount of graphite in the MML can reduce the shear strength of the near-surface region, thus reducing the COF and wear rate. By comparing uncoated and coated graphites, it is obvious that the coated graphite avoids bonding between graphite particles and reduces their clustering. Consequently, graphite phase has smaller size and average distance between particles. For this reason, the COF of coated composites decreases significantly at lower volume fraction of graphite in comparison with uncoated composites.

The involved wear mechanisms of sintered copper, Cu-coated, and uncoated graphite composites could be deduced from studying the microstructural changes at the surface and subsurface of tested pins at each transition load regime. Generally, the wear mechanism of pure copper and Cu/graphite composites at different applied loads was described by Moustafa et al. [69], and the situation of the wear mechanism is different for Cu/graphite composites due to the presence of graphite. Pure copper and Cu/graphite detects very low plastic deformation within the structure at low wear regime. Certain amount of Cu_2O is observed on the copper surface (Figure 2.24a) due to atmosphere oxidation of copper, whereas the surface of composites contains graphite particles as well as Cu_2O (Figure 2.25a). This layer is a loosely adherent film. Accordingly, the oxidative wear is the dominant wear mechanism with very fine identical particles as wear debris. Very few debris are generated consisting of fine equiaxed particles for copper composites (Figure 2.26a). The investigation of collected debris from copper samples is shown in Figure 2.27. The formed debris become metallic flakes of sharp edges. At the mild wear regime, a considerable deformation of the structure is observed beneath the wear surface (Figure 2.24b), whereas the surface of the composite is covered with a graphite lubricant film (40–200 µm; Figure 2.25b). A mixed layer is observed at the surface and subsurface of the sample, and larger wear debris are formed. Compacted particle-like copper and cracked-shaped graphite wear debris are formed for copper composites (Figure 2.27b). Severe wear regime shows extensive deformation, fragmentation, and several grooves at the contacting surface and subsurface (Figures 2.24c and 2.25c). Predominantly large and sharp-edged debris are generated in both cases of copper composites (Figure 2.27c).

Figure 2.28 shows the effect of the load on the COF of copper and its composites. It is clearly confirmed that the COF gradually decreases with increasing loads up to a load of 30 N with increased. At higher applied load, the COF of all the composites gradually remains almost constant. It occurs due to the fact that the copper matrix becomes soft and highly ductile under severe deformation. Consequently, it results in easy transfer of copper metallic film on the counter disc, and this film prevents the direct contact between the specimen and the counter disc leading to lowering of COF.

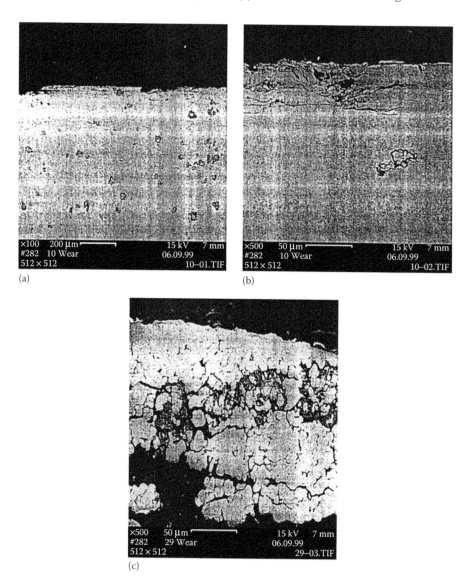

FIGURE 2.24 Microstructural changes due to wear testing at the surface and subsurface of sintered copper compacts at (a) low, (b) mild, and (c) severe wear regimes. (From Moustafa, S. et al., *Wear*, 253, 699–710, 2002.)

Additionally, copper/graphite composites exhibit the lowest COF, and copper/silicon carbide composite exhibits the highest, whereas Cu–SiC–graphite composite shows an intermediate value.

The presence of graphite reinforcement has a negative effect on the mechanical properties of copper composites. The best choice is employing nanosized particles to solve the impact of microsized graphite particles. Rajkumar et al. [70]

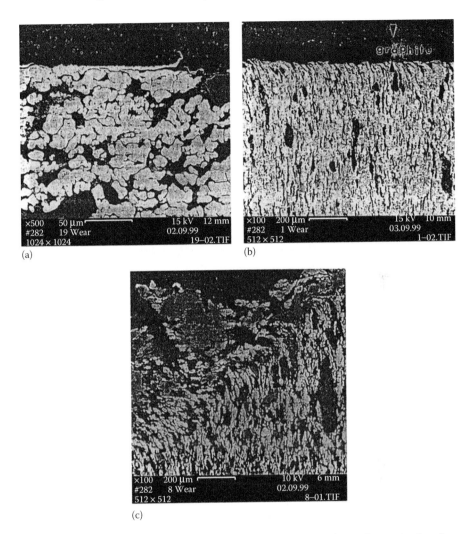

FIGURE 2.25 Microstructural changes due to wear testing at the surface and subsurface of either Cu-coated or uncoated graphite composites at (a) low, (b) mild, and (c) severe wear regimes. (From Moustafa, S. et al., *Wear*, 253, 699–710, 2002.)

have investigated the tribological properties of copper nanocomposite reinforced by 5–20 vol.% nano-graphite particles with an average particle size of 35 nm. The variation of normal load with COF at different volume fractions of micro- and nano-graphite particles is shown in Figure 2.29. Figure 2.30 shows the variation of wear rate with normal load for copper-based composites reinforced by micro- and nano-graphite particles. It can be inferred from Figures 2.29 and 2.30 that the wear rate and COF both increase with increasing applied load. The results revealed that, at constant 15% volume fraction, nanocomposites have better COF compared to the composite reinforced by micron-sized graphite particles due to higher hardness,

Probe 11b   100 μm
(a)

Probe 11b   100 μm
(b)

Probe 1000   10 mm
(c)

FIGURE 2.26 Micrographs of wear debris generated from sintered copper compacts at (a) low, (b) mild, and (c) severe wear regimes. (From Moustafa, S. et al., *Wear*, 253, 699–710, 2002.)

lower porosity, and finer microstructure. Therefore, the composites reinforced by nano-graphite particles are more effective to the degree of self-lubrication compared to the composites reinforced by micron-size graphite particles. In addition, the volume fraction of nano-graphite can affect the tribological properties of self-lubricating copper composites. Increasing the amount of nano-graphite decreases the COF up to 20%, and after that, the COF increases due to the formation of a uniform and continuous layer of the solid lubricant film. The reason for the decrease of COF and wear rate at high volume fraction of nano-graphite is the formation

Probe 11c23_b
(a)

Probe 23P   300 μm
(b)

Probe 8 m   300 μm
(c)

FIGURE 2.27 Typical example of microstructural changes of wear debris generated from either Cu-coated or uncoated graphite composites at (a) low, (b) mild, and (c) severe wear regimes. (From Moustafa, S. et al., *Wear*, 253, 699–710, 2002.)

of a lubricant film with more availability and uniformity. In general, the lubricant film tends to decrease the metal-to-metal contacts between the copper matrix composite and the steel counter surface. Conversely, at higher amount of nano-graphite (20 vol.%), agglomeration occurs, and it leads to incomplete spreading of graphite at the contact zone, and thus increases the COF and the wear rate.

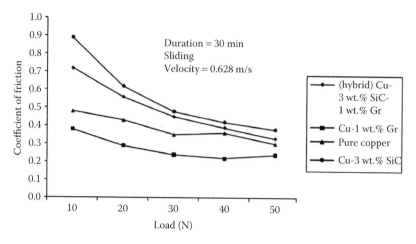

FIGURE 2.28 Variation of COF of copper and copper–SiC–graphite with load. (From Ramesh, C. et al., *Mater. Des.*, 30, 1957–1965, 2009.)

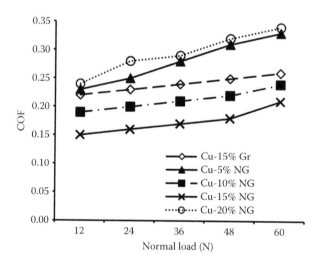

FIGURE 2.29 Variation of COF with normal load at a sliding speed of 0.77 m/s and variation of COF with sliding speed at 36 N. (From Rajkumar, K. and Aravindan, S., *Tribol. Inter.*, 57, 282, 2013.)

Some researchers have studied the influence of different solid lubricant particles on the tribological behavior of copper matrix composites. For instance, Chen et al. [75] have investigated the tribological properties of two different solid lubricants, graphite and hBN, for copper composites. Copper matrix composites reinforced by graphite at weight fractions of 0%, 2%, 5%, 8%, and 10%, corresponding to the hBN

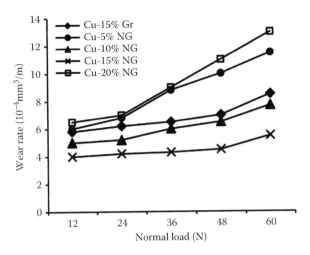

FIGURE 2.30 Variation of the wear rate of composites with normal load at a sliding speed of 0.77 m/s. (From Rajkumar, K. and Aravindan, S., *Tribol. Inter.*, 57, 282, 2013.)

at weight fractions of 10%, 8%, 5%, 2%, and 0%, respectively. Figure 2.31 shows the variation of COF and wear rate at different normal loads. The results show that lubrication effects of graphite are superior to those of hBN. Also, composites with higher amount of graphite significantly show better wear rates and friction coefficient.

Figure 2.32 shows the crystalline structures of graphite and hBN [76]. C–C bonds in graphite and B–N bonds in hBN can be observed between neighbor interlayers of the crystalline structure. Strong electrical dipoles between B and N atoms make a strong van der Waals force between interlayers of h-BN compared to the graphite layer. Therefore, the interplanar spacing of hBN is shorter than that of graphite where the distances between the neighbor interlayers of hBN and graphite are 3.33 Å and 3.35 Å [77], respectively. This shorter interplanar spacing resulted in a strong bonding in the crystalline structure of hBN, which could tend to different lubrication effects on copper-based composites. Weaker interlayer bonding of graphite gives this ability to easily shear along the basal plane of the crystalline structures than that of hBN. Accordingly, the wear rates of the sample with 8 and 10 vol.% of graphite are much lower than those of others due to the formation of compact and continuously tribo-films of the samples with 8% and 10% graphite (Figure 2.33). In addition, sheared particles of graphite are finer than those of hBN in the wear process (Figure 2.34).

Kato et al. [78] have investigated that the wear rates of Cu–MoS_2 composites increased considerably with the MoS_2 addition. Because there is a considerable reduction in the hardness of Cu–Gr composites, an extensive transfer of copper (which is highly ductile in nature) onto the counter steel disc surfaces of very high hardness of RC-60 occurs. This phenomenon has been observed by Kestursatya et al. [79] for Cu–Gr composites.

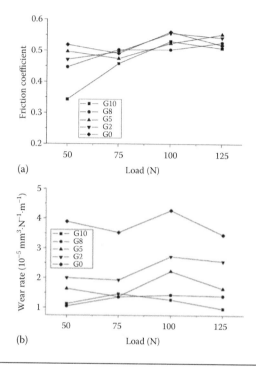

(a)

(b)

Sample Designation	Graphite (30 μm; 99% Purity)	BN (0.65–11.38 μm; 99% Purity)	SiC (75–150 μm; 98.5% Purity)	Sn+Al+Fe (75–150 μm; 98% Purity)
G0	0	10	6	18
G2	2	8	6	18
G5	5	5	6	18
G8	8	2	6	18
G10	10	0	6	18

FIGURE 2.31   Variation of (a) friction coefficient and (b) wear rates with loads for various copper-based composites. (From Chen, B. et al., *Tribol. Inter.*, 41, 1145–1152, 2008.)

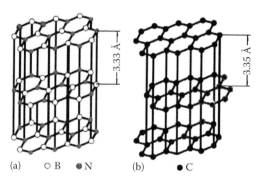

(a) ○ B ● N   (b)   ● C

FIGURE 2.32   Crystalline structures of (a) hBN and (b) graphite. (From Rohatgi, P.K. et al., *Inter. Mater. Rev.*, 37, 129–152, 1992.)

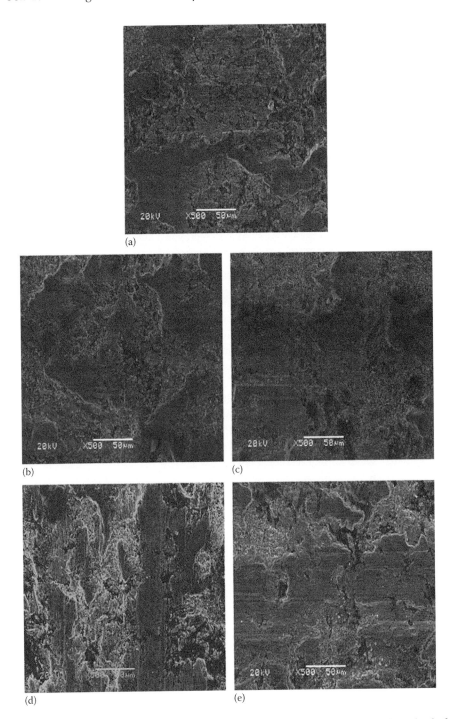

FIGURE 2.33   SEM micrographs of the worn surfaces of the composites tested at a load of 100 N and a speed of 2.6 m/s: (a) G0, (b) G2, (c) G5, (d) G8, and (e) G10.

(a)   (b)

FIGURE 2.34 Wear debris of the Cu-based composites tested at a load of 100 N and a sliding speed of 2.6 m/s: (a) G2 and (b) G10. All bars represent 5 mm.

2.4 MAGNESIUM MATRIX COMPOSITES

Magnesium alloys have been used in automobile and aerospace industries due to their low density, high specific strength and stiffness, good damping characteristic, excellent machineability, and castability. However, its corrosion [80] and wear resistance [81] limit it to be used as widely as aluminum alloy. The most commonly used magnesium alloy is AZ91 alloy. Qi [82] studied the influence of graphite particle content on the friction and wear characteristics of AZ91 magnesium alloy matrix composite. With regard to the wear rate (Figure 2.35), the wear behavior of composites containing different contents of graphite particles shows the wear mass loss of the composites as well as the base alloy specimens as a function of the applied load. The results showed that the wear resistances of graphite containing composite are much better than those of the matrix under similar testing conditions. The wear mass loss of each composite specimen decreases with the increase in graphite content. The authors reported that

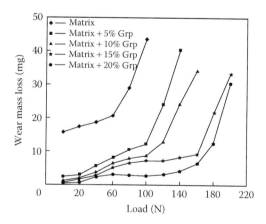

FIGURE 2.35 Relation between mass loss and load. (From Qi, Q.-J., *Transac. Nonferr. Met. Soc. China*, 16, 1135–1140, 2006.)

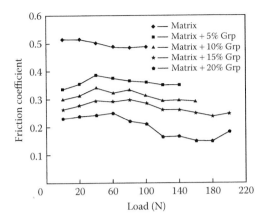

FIGURE 2.36 Relation between friction coefficient and load. (From Qi, Q.-J., *Transac. Nonferr. Met. Soc. China*, 16, 1135–1140, 2006.)

a continuous black lubricating film forms progressively on the worn surface during sliding, which effectively limits the direct interaction between the composite tribo-surface and the counterpart, and also remarkably delays the transition from mild wear to severe wear for magnesium alloy composites. It can be seen in Figure 2.36 that the friction coefficients of composites are much lower than that of the matrix alloy. The graphite content presents a significant effect on the friction coefficient of the specimens, and the friction coefficients decrease with increasing graphite content.

It can be seen from the results that at lower loads, comparatively low wear rates exist in the regime of mild wear. In Figure 2.37a, EDX analysis shows the presence of the elements of O (39.98 wt.%), Mg (41.42 wt.%), Fe (15.91 wt.%), and Al (3.70 wt.%), implying that the surface layer consists of the oxide mixture of these elements. With the increase of applied loads, the subsurface of the specimens is subjected to more severe damage. At 40 N normal load, fragmentation of the reinforcement induces abrasive wear, manifesting itself by distinct scratches and grooves on the worn surface (Figure 2.37b). Consequently, it can be concluded that the main wear mechanism in this process is abrasive wear. Black lubricating film forms progressively on the sliding surface under the normal loads of 60–160 N, as revealed in Figure 2.37c–e. Graphite particles are apt to be covered by plastic flow of the matrix and the debris under lower loads, which inhibits graphite to spread sufficiently. It can be seen from Figure 2.37c that only part of the worn surface possesses black lubricating film. In Figure 2.37d and e, a continuous lubricating film almost covers the whole sliding surface. The formation of the film of graphite during sliding limits effectively the direct metal-to-metal interaction between the composite tribo-surface and the counterpart, and changes the relationship among friction coefficient, wear mass loss, and load, corresponding to the lowest friction coefficient and the sluggish increase in the wear rate. At the normal load of 10 N, the pin surface is subjected to light ploughing and no obvious grooves can be found. The wear debris are very fine and small in size as shown in Figure 2.38a, which gives rise to a relatively low wear mass loss.

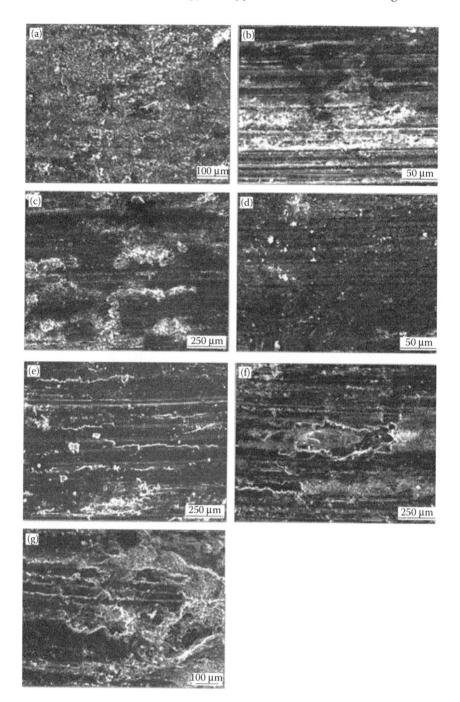

FIGURE 2.37 Morphology of worn surfaces of 15% graphite particle-reinforced composites under different loads: (a) 10 N, (b) 40 N, (c) 60 N, (d) 100 N, (e) 160 N, (f) 160 N, and (g) 200 N. (From Qi, Q.-J., *Transac. Nonferr. Met. Soc. China*, 16, 1135–1140, 2006.)

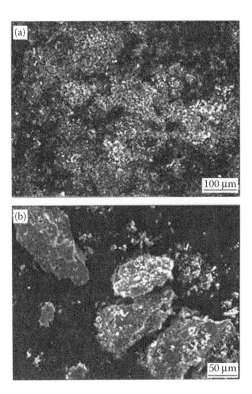

FIGURE 2.38 Debris morphologies of 15% graphite particle-reinforced composite: (a) 10 N and (b) 160 N.

Zhang et al. [83] studied the effect of graphite particle size on the wear property of graphite and Al_2O_3 reinforced AZ91D-0.8%Ce composites. Figure 2.39 shows the variation of wear loss with load. It can be observed that the embedded graphite in the matrix acts as a lubricant and decreases the wear loss. The wear resistance of the composites increases as the graphite particle size increases. At low load,

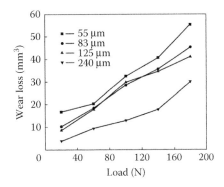

FIGURE 2.39 Variations of wear loss with load of composites. (From Zhang, M.-J. et al., *Transac. Nonferr. Met. Soc. China*, 18, s273–s277, 2008.)

the composites have similar wear loss; at high load, the composite with the largest graphite particle size has the best wear resistance. The wear mechanism of all the composites at low load is abrasive wear and oxidation wear; at high load, the wear mechanism of the composites changes to delamination wear. The composite with 240 μm particle has evidently bare graphite particle on the worn surface, as pointed in Figure 2.40a. At 100 N, the oxidation phenomenon of the composite with 55 μm particle deteriorates, but the worn surface of the composite with 240 μm particle changes less. And the wear mechanism of both is still abrasive wear and oxidation

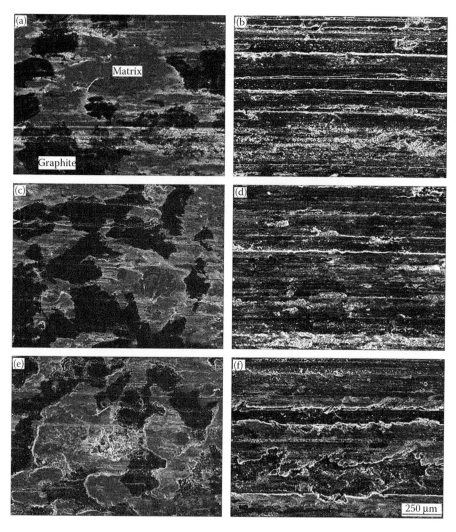

FIGURE 2.40 SEM morphologies of the worn surface of composites with a particle size of 240 μm at 20 N (a), 100 N (c), 180 N (e) and with a particle size of 55 μm at 20 N (b), 100 N (d), and 180 N (f). (From Zhang, M.-J. et al., *Transac. Nonferr. Met. Soc. China*, 18, s273–s277, 2008.)

wear. This reveals that no matter how the graphite particle size is larger or small, the presence of graphite can always keep the composites with high wear resistance. When the tested load increases up to 180 N, the composite with 55 μm particle has large flakes peeled along the sliding direction combining with the appearance of cracks. The graphite particle cannot keep intact, and it is peeled off with flakes. The wear loss increases sharply, which accords with the curve in Figure 2.39. Thus, its wear mechanism transforms to delamination wear.

However, the worn surface of the composite with 240 μm particle looks the same as that under 100 N, except that the partial area of the matrix is peeled off. This is because the composite with 240 μm particle has evident bare graphite particle on the worn surface. At low load, the graphite particle decreases the actual contact area between composite matrix and friction disk; at high load, it is extruded and smeared on the worn surface and benefits to form graphite film [84,85]. The graphite film has a lubricious effect and descends the extent of the friction. Finally, the presence of graphite film decreases the wear loss of the composite. Therefore, the wear mechanism of the composite with the 240 μm particle is still abrasive wear and oxidation wear.

Mindivan et al. [86] investigated the effect of nano solid lubricant on the tribological properties of magnesium self-lubricating materials. When 0.5 wt.% CNTs embedded into the magnesium matrix, significant improvements in the COF and wear rate were observed. Further, increasing the amount of CNTs reduces the wear rate and COF as shown in Figure 2.29. The maximum effective content of CNTs is 2 and 4 wt.% where a reduction in the wear rate and COF is found to be the highest (Figure 2.41). Figure 2.42 presents the SEM and two-dimensional (2D) profile images of the wear tracks formed on the base alloy and composite containing 2 wt.% CNTs. The morphology of the worn surface of base alloy is rough, whereas the composite with 2 wt.% CNTs showed a relatively smoother appearance (Figure 2.42). The depth of the wear track of the composite containing 2 wt.% CNT is less than that of the base alloy. With the addition of CNTs, a notable reduction in the presence of craters and the formation of thin and adherent transfer film on

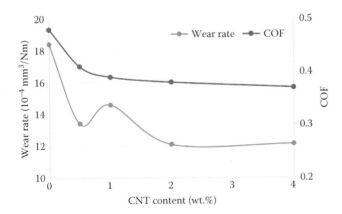

FIGURE 2.41 Effect of CNTs on the wear rate and COF of magnesium nanocomposites. (From Mindivan, H. et al., *Appl. Surf. Sci.*, 318, 234–243, 2014.)

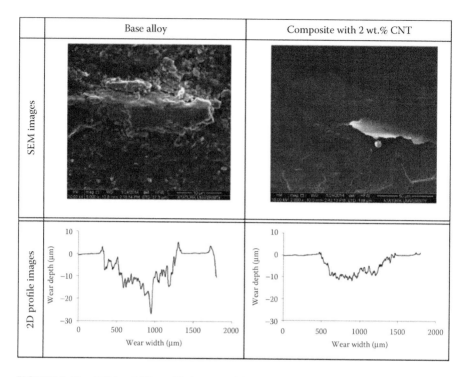

FIGURE 2.42 SEM and 2D profile images of the wear tracks formed on the base alloy and composite containing 2 wt.% CNT. (From Mindivan, H. et al., *Appl. Surf. Sci.*, 318, 234–243, 2014.)

the worn surface were observed (Figure 2.42). Hence, the enhancement of the wear properties may come from the presence of CNTs that act as a lubricant medium reducing the COF.

2.5 NICKEL MATRIX COMPOSITES

Excellent high-temperature performance of nickel–graphite composites makes some application of these composites for high-efficiency engine. Some studies have been investigating the effect of materials and test parameters on the tribological behavior of nickel–graphite composites. Li et al. [87] studied the effects of the tribological temperature, load, and speed on the wear behavior of nickel–graphite composite. The variation of COF and wear rates of nickel–graphite composite as a function of the graphite volume fraction, load, and speed is shown in Figure 2.43. Addition of graphite particles shows a significant improvement in the friction and wear properties rather than pure nickel, and 6–12 wt.% of graphite is the optimum range of graphite content where the COF and wear rate are minimized. As shown in Figure 2.43,

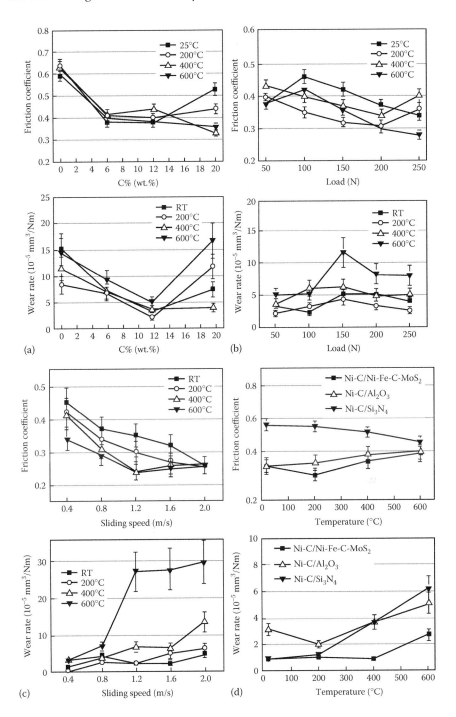

FIGURE 2.43 Variation of COF and wear rates for different (a) graphite content, (b) load, (c) speed, and (d) counterpart materials. (From Li, J. and Xiong, D., *Wear*, 266, 360–367, 2009.)

increasing the load and sliding speed reduces friction coefficients, whereas the wear rates increase with the increasing temperature and sliding speed. In addition, the wear rate increases by increasing the normal load up to 150 N and then decreases by increasing load.

Further, Li et al. [88] studied the tribological properties of Ni–Cr–W–Fe–C self-lubricating composite reinforced by graphite and MoS_2 as a solid lubricant at different temperatures. It was found that chromium sulfide and tungsten carbide were formed in the composite by adding MoS_2 and graphite, which were responsible for low friction and high wear resistance at elevated temperatures, respectively. Figure 2.44 shows the variation of COF and wear rate for different composites at different temperature. Amongst the composites, nickel composites reinforced by graphite and MoS_2 exhibits better tribological properties for a wide range of temperatures due to the synergistic lubricating effect of graphite and MoS_2. The graphite played the main

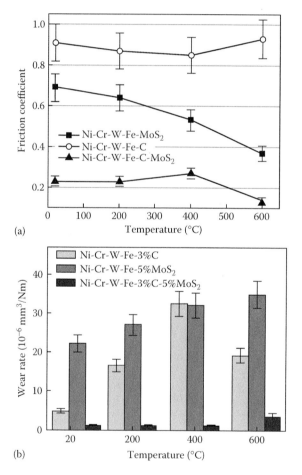

FIGURE 2.44 Variation of (a) COF and (b) wear rate in the composites reinforced by graphite and MoS_2. (From Li, J.L. and Xiong, D.S., *Wear*, 265, 533–539, 2008.)

role of lubrication at room temperature, whereas sulfides were responsible for low friction at high temperature.

Chen et al. [89] and Scharf et al. [90] investigated the effect of CNTs on the tribological behavior of self-lubricating nickel composites. The results show a reduction in COF by embedding CNT nanoparticles. In addition, Scharf et al. [90] found that adding CNTs is more efficient than adding graphite microparticles due to strong mechanical properties of CNTs. CNTs comprise concentric cylindrical layers or shells of graphite-like sp^2-bonded cylindrical layers or shells, where the intershell interactions are predominately van der Waals and can easily slide or rotate each other, leading to a low friction coefficient. Figure 2.45 shows the variation of friction coefficients and the wear rate with load at several concentrations of CNTs. The COF results reveal that increasing the load decreases COF, whereas the wear rate increases by increasing the load. Also, the optimum concentration of CNTs is

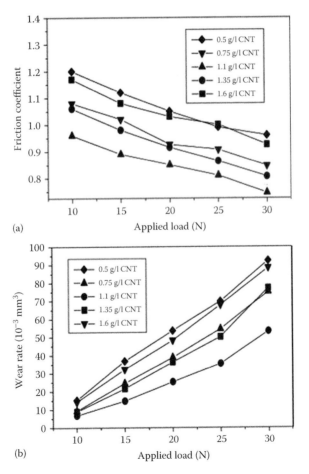

FIGURE 2.45 Effect of CNT concentration on (a) COF and (b) wear rate of Ni/CNT composites. (From Scharf, T. et al., *J. Appl. Phys.*, 106, 013508, 2009.)

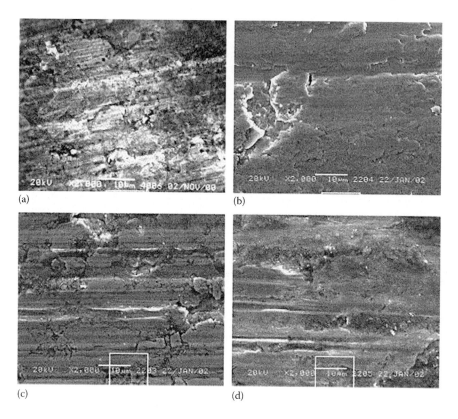

FIGURE 2.46 Wear morphology of (a) Ni, (b) Ni/SiC, (c) Ni/graphite, and (d) Ni/CNT coatings. (From Scharf, T. et al., *J. Appl. Phys.*, 106, 013508, 2009.)

1.1 g/L where the COF and wear rate possess minimum value. Figure 2.46 compares the worn surface of Ni, Ni/graphite, and Ni/CNT composites. Large grooves and a considerable extent of peeling on the worn surface of the Ni are present. By comparing the worn surface of nickel composites reinforced by graphite and CNTs, it reveals that there are less cracks and finer scratches on the surface of nickel/CNTs rather than nickel/graphite, resulting in a less amount of material loss by embedding CNTs.

REFERENCES

1. Daniel IM, Ishai O. *Engineering Mechanics of Composite Materials.* Oxford University Press: New York; 1994.
2. Prasad S, Asthana R. Aluminum metal-matrix composites for automotive applications: Tribological considerations. *Tribology Letters.* 2004;17(3):445–453.
3. Miracle DB. Metal matrix composites—From science to technological significance. *Composites Science and Technology.* 2005;65(15–16):2526–2540.

4. Omrani E et al. Influences of graphite reinforcement on the tribological properties of self-lubricating aluminum matrix composites for green tribology, sustainability, and energy efficiency—a review. *The International Journal of Advanced Manufacturing Technology*. 2016;83(1–4):325–346.

5. Omrani E et al. New emerging self-lubricating metal matrix composites for tribological applications. In: Davim PJ, (Ed). *Ecotribology: Research Developments*. Springer International Publishing: Cham, Switzerland; 2016. pp. 63–103.

6. Dellacorte C, Fellenstein JA. The effect of compositional tailoring on the thermal expansion and tribological properties of PS300: A solid lubricant composite coating. *Tribology Transactions*. 1997;40(4):639–642.

7. Zhang X et al. Carbon nanotube-MoS_2 composites as solid lubricants. *ACS Applied Materials and Interfaces*. 2009;(3):735–739.

8. Chromik RR et al. In situ tribometry of solid lubricant nanocomposite coatings. *Wear*. 2007;262(9–10):1239–1252.

9. Menezes PL, Rohatgi PK, Lovell MR. Self-lubricating behavior of graphite reinforced metal matrix composites. In: Nosonovsky M, Bhushan B, (Eds.). *Green Tribology*. Springer: Berlin, Germany; 2012, pp. 445–480.

10. Moghadam AD et al. Mechanical and tribological properties of self-lubricating metal matrix nanocomposites reinforced by carbon nanotubes (CNTs) and graphene–A review. *Composites Part B: Engineering*. 2015;77:402–420.

11. Rohatgi PK. Metal matrix composites. *Defence Science Journal*. 1993;43(4):323.

12. Liu Y, Rohatgi PK, Ray S. Tribological characteristics of aluminum-50 Vol Pct graphite composite. *Metallurgical Transactions A*. 1993;24(1):151–159.

13. Rohatgi PK, Ray S, Liu Y. Tribological properties of metal matrix-graphite particle composites. *International Materials Reviews*. 1992;37:129–152.

14. Rohatgi PK et al. A surface-analytical study of tribodeformed aluminum alloy 319–10 vol.% graphite particle composite. *Materials Science and Engineering: A*. 1990;123(2):213–218.

15. Jha A et al. Aluminium alloy-solid lubricant talc particle composites. *Journal of Materials Science*. 1986;21(10):3681–3685.

16. Bowden F, Shooter K. Frictional behaviour of plastics impregnated with molybdenum disulphide. *Industrial & Engineering Chemistry Research*. 1950;3:384.

17. Lancaster J. Composite self-lubricating bearing materials. *Proceedings of the Institution of Mechanical Engineers*. 1967;182(1):33–54.

18. Prasad S, Mecklenburg KR. Self-lubricating aluminum metal-matrix composites dispersed with tungsten disulfide and silicon carbide. *Lubrication Engineering*. 1994;50(7). https://www.osti.gov/scitech/biblio/191823.

19. Skeldon P, Wang H, Thompson G. Formation and characterization of self-lubricating MoS_2 precursor films on anodized aluminium. *Wear*. 1997;206(1):187–196.

20. Rawal SP. Metal-matrix composites for space applications. *JOM*. 2001;53(4):14–17.

21. Reboul M, Baroux B. Metallurgical aspects of corrosion resistance of aluminium alloys. *Materials and Corrosion*. 2011;62(3):215–233.

22. Molina J-M et al. Thermal conductivity of aluminum matrix composites reinforced with mixtures of diamond and SiC particles. *Scripta Materialia*. 2008;58(5):393–396.

23. Recoules V et al. Electrical conductivity of hot expanded aluminum: Experimental measurements and ab initio calculations. *Physical Review E*. 2002;66(5):056412.

24. Li G-C et al. Damping capacity of high strength-damping aluminum alloys prepared by rapid solidification and powder metallurgy process. *Transactions of Nonferrous Metals Society of China*. 2012;22(5):1112–1117.

25. Moghadam AD, Schultz BF, Ferguson JB, Omrani E, Rohatgi PK, Gupta N. Functional metal matrix composites: Self-lubricating, self-healing, and nanocomposites-an outlook. *JOM*. 2014;66(6):872–81.

26. Kumar GV, Rao C, Selvaraj N. Mechanical and tribological behavior of particulate reinforced aluminum metal matrix composites—a review. *Journal of Minerals and Materials Characterization and Engineering*. 2011;10(1):59.

27. Kathiresan M, Sornakumar T. Friction and wear studies of die cast aluminum alloy-aluminum oxide-reinforced composites. *Industrial Lubrication and Tribology*. 2010;62(6):361–371.

28. Rohatgi P, Ray S, Liu Y. Tribological properties of metal matrix-graphite particle composites. *International Materials Reviews*. 1992;37:129–152.

29. Ames W, Alpas A. Wear mechanisms in hybrid composites of Graphite-20 Pct SiC in A356 aluminum alloy (Al-7 Pct Si-0.3 Pct Mg). *Metallurgical and Materials Transaction A*. 1995;26(1):85–98.

30. Roy M et al. The effect of participate reinforcement on the sliding wear behavior of aluminum matrix composites. *Metallurgical Transactions A*. 1992;23(10):2833–2847.

31. Alpas A, Zhang J. Effect of microstructure (particulate size and volume fraction) and counterface material on the sliding wear resistance of particulate-reinforced aluminum matrix composites. *Metallurgical and Materials Transactions A*. 1994;25(5):969–983.

32. Van AK et al. Influence of tungsten carbide particle size and distribution on the wear resistance of laser clad WC/Ni coatings. *Wear*. 2005;258(1):194–202.

33. Nayeb-Hashemi H, Seyyedi J. Study of the interface and its effect on mechanical properties of continuous graphite fiber-reinforced 201 aluminum. *Metallurgical Transactions A*. 1989;20(4):727–739.

34. Tokisue H, Abbaschian G. Friction and wear properties of aluminum-particulate graphite composites. *Materials Science and Engineering*. 1978;34(1):75–78.

35. Ravindran P et al. Investigation of microstructure and mechanical properties of aluminum hybrid nano-composites with the additions of solid lubricant. *Materials and Design*. 2013;51:448–456.

36. Shanmughasundaram P, Subramanian R. Wear behaviour of eutectic Al-Si alloy-graphite composites fabricated by combined modified two-stage stir casting and squeeze casting methods. *Advances in Materials Science and Engineering*. Article ID 216536, 2013; 8. http://dx.doi.org/10.1155/2013/216536.

37. Ravindran P et al. Tribological behaviour of powder metallurgy-processed aluminium hybrid composites with the addition of graphite solid lubricant. *Ceramics International*. 2013;39(2):1169–1182.

38. Suresha S, Sridhara BK. Wear characteristics of hybrid aluminium matrix composites reinforced with graphite and silicon carbide particulates. *Composites Science and Technology*. 2010;70(11):1652–1659.

39. Srivastava S et al. Study of the wear and friction behavior of immiscible as cast-Al-Sn/Graphite composite. *International Journal of Modern Engineering Research*. 2012;2(2):25–42.

40. Akhlaghi F, Pelaseyyed SA. Characterization of aluminum/graphite particulate composites synthesized using a novel method termed "in-situ powder metallurgy". *Materials Science and Engineering*: A. 2004;385(1–2):258–266.

41. Hocheng H et al. Fundamental turning characteristics of a tribology-favored graphite/aluminum alloy composite material. *Composites Part A: Applied Science and Manufacturing*. 1997;28(9–10):883–890.

42. Baradeswaran A, Perumal E. Wear and mechanical characteristics of Al 7075/graphite composites. *Composites: Part B*. 2014;56:472–476.

43. Akhlaghi F, Zare-Bidaki A. Influence of graphite content on the dry sliding and oil impregnated sliding wear behavior of Al 2024–graphite composites produced by in situ powder metallurgy method. *Wear*. 2009;266(1–2):37–45.

44. Akhlaghi F, Mahdavi S. Effect of the SiC content on the tribological properties of hybrid Al/Gr/SiC composites processed by in situ powder metallurgy (IPM) method. *Advanced Materials Research*. 2011;264–265:1878–1886.

45. Baradeswaran A, Perumal AE. Study on mechanical and wear properties of Al 7075/Al2O3/graphite hybrid composites. *Composites: Part B*. 2014;56:464–471.

46. Basavarajappa S et al. Dry sliding wear behavior of Al 2219/SiCp-Gr hybrid metal matrix composites. *Journal of Materials Engineering and Performance*. 2006;15(6):668–674.

47. Prasad BK, Das S. The signifance of the matrix microstructure on the solid lubrication characterstics in aluminum alloys. *Materials Science and Engineering A*. 1991;144:229–235.

48. Radhika N et al. Dry sliding wear behaviour of aluminium/alumina/graphite hybrid metal matrix composites. *Industrial Lubrication and Tribology*. 2012;64(6):359–366.

49. Babić M et al. Wear properties of A356/10SiC/1Gr hybrid composites in lubricated sliding conditions. *Tribology in Industry*. 2013;35(2):148–154.

50. Baradeswaran A, Elayaperumal A. Wear characteresitic of Al6061 reinforced with graphite under different loads and speeds. *Advanced Materials Research*. 2011;287–290:998–1002.

51. Rajaram G, Kumaran S, Rao TS. Fabrication of Al–Si/graphite composites and their structure–property correlation. *Journal of Composite Materials*. 2011;45(26):2743–2750.

52. Chen Z et al. Microstructure and properties of in situ Al/TiB2 composite fabricated by in-melt reaction method. *Metallurgical and Materials Transactions A*. 2000;31(8):1959–1964.

53. Tjong SC. Novel Nanoparticle-reinforced metal matrix composites with enhanced mechanical properties. *Advanced Engineering Materials*. 2007;9(8):639–652.

54. Thostenson ET, Li C, Chou T-W. Nanocomposites in context. *Composites Science and Technology*. 2005;65(3):491–516.

55. He F, Han Q, Jackson MJ. Nanoparticulate reinforced metal matrix nanocomposites—a review. *International Journal of Nanoparticles*. 2008;1(4):301–309.

56. Singh J, Narang D, Batra NK. Experimental investigation of mechanical and tribological properties of Aa-SiC and Al-Gr metal matrix composite. *International Journal of Engineering Science and Technology*. 2013;5(6):1205–1210.

57. Ghasemi-Kahrizsangi A, Kashani-Bozorg SF. Microstructure and mechanical properties of steel/TiC nano-composite surface layer produced by friction stir processing. *Surface & Coatings Technology*. 2012;209:15–22.

58. Tabandeh-Khorshid M, Jenabali-Jahromi SA, Moshksar MM. Mechanical properties of tri-modal Al matrix composites reinforced by nano- and submicron-sized Al_2O_3 particulates developed by wet attrition milling and hot extrusion. *Materials & Design*. 2010;31(8):3880–3884.

59. Shafiei-Zarghani A, Kashani-Bozorg SF, Zarei-Hanzaki A. Microstructures and mechanical properties of Al/Al2O3 surface nano-composite layer produced by friction stir processing. *Materials Science and Engineering A*. 2009;500:87–91.

60. Jinfeng L et al. Effect of graphite particle reinforcment on dry sliding wear of SiC/Gr/Al composites. *Rare Metal Materials and Engineering*. 2009;38(11):1894–1898.

61. Rohatgi PK et al. Tribology of metal matrix composites. In: Menezes P, Ingole SP, Nosonovsky M, Kailas SV, Lovell MR, (Eds.). *Tribology for scientists and engineers*. Springer: New York; 2013. pp. 233–268.

62. Bakshi S, Lahiri D, Agarwal A. Carbon nanotube reinforced metal matrix composites—A review. *International Materials Reviews*. 2010;55(1):41–64.

63. Kovacik J et al. Effect of composition on friction coefficient of Cu–graphite composites. *Wear*. 2008;256:417–421.

64. Zhou S-M et al. Fabrication and tribological properties of carbon nanotubes reinforced Al composites prepared by pressureless infiltration technique. *Composites: Part A.* 2007;38(2):301.

65. Choi HJ, Lee SM, Bae DH. Wear characteristic of aluminum-based composites containing multi-walled carbon nanotubes. *Wear.* 2010;270(1–2):12.

66. Ghazaly A, Seif B, Salem HG. Mechanical and tribological properties of AA2124-graphene self lubricating nanocomposite. *Light Metals.* 2013;2013:411–415.

67. Zamzam M. Wear resistance of agglomerated and dispersed solid lubricants in aluminium. *Materials Transactions, JIM.* 1989;30(7):516–522.

68. Dharmalingam S et al. Optimization of tribological properties in aluminum hybrid metal matrix composites using gray-Taguchi method. *Journal of Materials Engineering and Performance.* 2011;20(8):1457–1466.

69. Moustafa S et al. Friction and wear of copper–graphite composites made with Cu-coated and uncoated graphite powders. *Wear.* 2002;253(7):699–710.

70. Zhao H et al. Investigation on wear and corrosion behavior of Cu–graphite composites prepared by electroforming. *Composites Science and Technology.* 2007;67(6):1210–1217.

71. Jincheng X et al. Effects of some factors on the tribological properties of the short carbon fiber-reinforced copper composite. *Materials & Design.* 2004;25(6):489–493.

72. Rohatgi P, Ray S, Liu Y. Tribological properties of metal matrix-graphite particle composites. *International Materials Reviews.* 1992;37(1):129–152.

73. Ramesh C et al. Development and performance analysis of novel cast copper–SiC–Gr hybrid composites. *Materials & Design.* 2009;30(6):1957–1965.

74. Rajkumar K, Aravindan S. Tribological behavior of microwave processed copper–nanographite composites. *Tribology International.* 2013;57:282.

75. Chen B et al. Tribological properties of solid lubricants (graphite, h-BN) for Cu-based P/M friction composites. *Tribology International.* 2008;41(12):1145–1152.

76. Petrescu M. Boron nitride theoretical hardness compared to carbon polymorphs. *Diamond and Related Materials.* 2004;13(10):1848–1853.

77. Hod O. Graphite and hexagonal boron-nitride have the same interlayer distance. Why?. *Journal of chemical theory and computation.* 2012;8(4):1360–1369.

78. Kato H et al. Wear and mechanical properties of sintered copper–tin composites containing graphite or molybdenum disulfide. *Wear.* 2003;255(1):573–578.

79. Kestursatya M, Kim J, Rohatgi P. Friction and wear behavior of a centrifugally cast lead-free copper alloy containing graphite particles. *Metallurgical and Materials Transactions A.* 2001;32(8):2115–2125.

80. Song G, Bowles AL, StJohn DH. Corrosion resistance of aged die cast magnesium alloy AZ91D. *Materials Science and Engineering: A.* 2004;366(1):74–86.

81. Chen H, Alpas A. Sliding wear map for the magnesium alloy Mg-9Al-0.9 Zn (AZ91). *Wear.* 2000;246(1):106–116.

82. Qi Q-J. Evaluation of sliding wear behavior of graphite particle-containing magnesium alloy composites. *Transactions of Nonferrous Metals Society of China.* 2006;16(5):1135–1140.

83. Zhang M-J et al. Effect of graphite particle size on wear property of graphite and Al_2O_3 reinforced AZ91D-0.8% Ce composites. *Transactions of Nonferrous Metals Society of China.* 2008;18:s273–s277.

84. Yang X-H et al. Microstructures and properties of graphite and Al_2O_3 short fibers reinforced Mg-Al-Zn alloy hybrid composites [J]. *Transactions of Nonferrous Metals Society of China.* 2006;16(s2):1–5.

85. Yong-bing L, Rohatgi P, Ray S. Tribological characteristics of aluminum-50% graphite composite [J]. *Metallurgical Transactions A*. 1993;24(1):151–159.
86. Mindivan H et al. Fabrication and characterization of carbon nanotube reinforced magnesium matrix composites. *Applied Surface Science*. 2014;318:234–243.
87. Li J, Xiong D. Tribological behavior of graphite-containing nickel-based composite as function of temperature, load and counterface. *Wear*. 2009;266(1):360–367.
88. Li JL, Xiong DS. Tribological properties of nickel-based self-lubricating composite at elevated temperature and counterface material selection. *Wear*. 2008;265(3):533–539.
89. Chen X et al. Dry friction and wear characteristics of nickel/carbon nanotube electroless composite deposits. *Tribology International*. 2006;39(1):22–28.
90. Scharf T et al. Self-lubricating carbon nanotube reinforced nickel matrix composites. *Journal of Applied Physics*. 2009;106(1):013508.

3 Self-Lubricating Polymer Matrix Composites

3.1 INTRODUCTION

Lubrication is critical to the operational safety and reliability of industrial manufacturing and processing. Lubrication technology has been widely used in industrial applications, including roller bearings, journal bearings, and gears. Efficient lubrication is valuable to dissipate frictional heat, extend fatigue life, and reduce friction and wear [1]. Existing lubrication systems rely on the use of synthetic lubrication oil or mineral oil, which cannot be used in strictly regulated fields such as pharmaceutical, food, and health-care industries due to the potential product contamination [2]. Solid lubricants are considered the best option to control friction and wear if the usage of liquid lubricants is not allowed. However, solid lubricants should meet certain requirements in practical applications, such as mechanical strength, stiffness, fatigue life, thermal expansion, and damping [3].

Certain polymer blends and composites achieve low friction coefficients ($\mu < 0.2$) and low wear rates ($k < 10^{-6}$ mm^3/Nm) without intentional external lubrication of the contact area [4–6]; as such, these materials are known as solid lubricants. Unfortunately, no general rules exist to guide the design of such materials, and discovery of solid lubricant materials almost always involves substantial trial and error. Depending on the relative properties of the filler and matrix, the filler can preferentially support load support [7], can arrest propagating cracks within the polymer [8], and can reduce shear strength of the sliding interface [9]. However, none of these functions explains the orders of magnitude effects that some small changes in filler composition [5,10], loading [5,11], or environment [12,13] are known to have on the wear rate of the system. Normally, the polymer produces wear debris during sliding against a hard metallic counterface, most commonly steel. These debris particles are dragged through the contact and ejected from the wear track in most cases. In special cases, the debris can adhere to the counterface to initiate the formation of a layer called the transfer film. Regardless of the materials used, a thin and continuous film is always observed on the steel counterpart following low wear sliding of polymeric materials. Likewise, a thick and patchy film is always observed after high wear sliding. Countless observations from the literature demonstrate a strong qualitative relationship between these properties of transfer films and measured wear rates [14–23].

As newly born polymer material, polymer composite exhibits excellent physical and chemical properties that include lightweight characteristics coupled with high strength, thermal stability, chemical resistance, flexural performance, relatively easy processing, self-lubricating capability, small friction coefficient, excellent anticorrosion property corrosion resistance, and outstanding wear resistance [24–26].

Polymer matrix composites (PMCs) are obvious candidates as self-lubricating substitutes for oil-lubricated materials. First, polymers generally have a low coefficient of friction (COF) (μ) during dry sliding, which excludes cold welding and consequently catastrophic sliding failure as seen for metals when lubricating systems fail [27]. Second, some polymers are intrinsically solid lubricants, for example, polytetrafluoroethylene (PTFE) [28,29]. Polymers are generally weak materials compared to metals and ceramics, which obviously results in some limitations. To overcome this drawback, different types of reinforcements, for example, fibers and particles, are often applied, which results in structured composite materials with excellent specific properties [30,31]. This is typically to improve mechanical properties but also in many cases for enhanced friction and wear control [32,33]. PMCs are increasingly utilized as sliding partners often against steel counterfaces, and relatively much empirical knowledge exists in this field. However, published data frequently seem to be conflicting and general guidelines are difficult to extract. This causes design and optimization of such tribological systems to be largely controlled by trial and error. Thus, there certainly is a need for more knowledge and understanding, which may lay the foundation for predictive guidelines for structuring self-lubricating PMCs toward enhanced friction and wear control.

Self-lubricating composites have been available for a long time and are used rather extensively by industry to combat friction and wear in a variety of sliding, rolling, and rotating bearing applications. They are generally prepared by dispersing appropriate amounts of a self-lubricating solid (as fillers, preferably in powder form) with a polymer. For example, it was shown that graphite, molybdenum disulfide (MoS$_2$), and boric acid fillers tend to increase the wear resistance of nylon and PTFE-type polymers [8,34]. Recent studies concluded that improved tribological behavior was mainly due to the formation of a thin transfer layer on the sliding surfaces of counterface materials. In the case of polymers, a significant increase in mechanical strength was also observed and thought to be responsible for high wear resistance. It was found that, initially, transfer films were not present but formed as a result of surface wear and subsurface deformation.

Solid lubrication has certain advantages compared to liquid lubrication. Basically, the solid lubricant can be provided as a coating on the tribological surface or inside a composite material. Solid lubrication has certain advantages compared to liquid lubrication. Basically, the solid lubricant can be provided as a coating on the tribological surface or inside a composite material.

3.2 EPOXY

Epoxies perhaps are the most favored matrices for developing high-performance composites especially for aircraft industries. In addition, they are used as high-grade synthetic resins, for example, in the electronics, aeronautics, and astronautics industries. Now epoxy is widely used in architecture, automotive, air, and railway transport systems for tribological applications. Epoxy resins are well established as thermosetting matrices of advanced composites, displaying a series of interesting characteristics, which can be adjusted within broad boundaries [35–38]. Being thermosets, these lack the advantage enjoyed by self-lubricating polymers such as

PTFE and high-density polyethylene, which are known for very low friction because of special feature of molecular chains sliding easily over each other. In recent years, not many papers have appeared on tribo-exploration of epoxies basically because of their limitation for high-temperature stability. Recently, AREMCO has commercialized an epoxy with high thermal stability (315°C), and hence, it was of interest to explore its potential as a matrix for developing high-performance tribo-composites.

Epoxy resins do typically exhibit relatively high wear rates and coefficients of friction when dry sliding against steel counterfaces. This is basically due to the cross-linked molecular structure, which inhibits the formation of an efficient transfer film and results in a relatively high degree of brittleness. However, epoxy resins possess other favorable properties such as strong adhesion to many materials, good mechanical and electrical properties, and relatively high chemical and thermal resistance.

For efficient tribo-composites apart from a thermally stable matrix, other requirements are of right reinforcement and solid lubricants. Among various reinforcements such as glass, aramid, graphite, and carbon, graphite has special features of very high thermal and thermo-oxidative stability, specific strength, thermal conductivity, and self-lubricity leading to an ideal choice for tribo-composites in spite of its high cost. Graphite fibers have the entire performance properties superior to carbon fibers (CFs). Comparatively less is reported on the exploration of GrF in tribology perhaps because of less availability and higher cost.

Because the CF is composed of many laminated graphite planes, the slip between the graphite layers easily occurs when subjected to shear traction, which gives the CF self-lubricating characteristics [27,39,40]. Therefore, CF-reinforced composites have been widely used in the bearings without liquid lubricant.

Sung and Suh [41] investigated the effect of fiber orientation on the friction and wear of CF-reinforced polymeric composites, and found that the friction and wear of composites were minimum when the fiber direction was perpendicular to the sliding surface. However, the composite surface with parallel fiber direction to the sliding surface is preferable because the exposed perpendicular CFs tend to gouge into the counter surface and initiate severe wear or seizure [18,42]. The minimum friction behavior of the CF-reinforced composite when the fiber direction was perpendicular to the sliding surface is contrary to intuition because the slip between the graphite layers in the CF occurs easily when the shear traction is parallel to the fiber direction, which may be caused by wear debris produced during sliding motion. Suh and Sin [43] investigated the increase of friction due to the wear debris and found that the friction increased exponentially as the plowing depth of wear debris increased.

A research compared the tribological properties of epoxy reinforced with glass and CFs at different normal loads and sliding speeds [44]. Figure 3.1 shows the measured values of COF for G/EP and CA/EP at different normal load and sliding velocity (p and v) conditions. The average level of COF for G/EP is 0.63 as opposed to 0.41 for CA/EP, which means that a general decrease of approximately 35% is obtained by substituting the glass fiber weave with the carbon/aramid weave. This difference in the level of COF might be attributed to the following factors: due to the hardness of glass fibers, these are observed by visual inspection to cause significant roughening and wear of the steel counterface at some conditions. Furthermore, fragments of glass fibers located in the interfacial zone can act as abrasive particles.

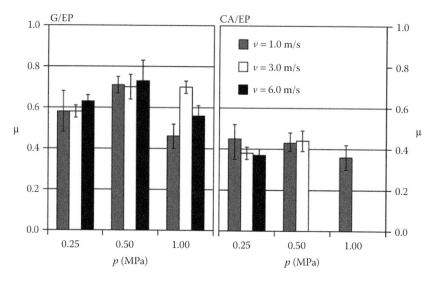

FIGURE 3.1 Coefficients of friction μ measured at different pv conditions for G/EP and CA/EP. EP, epoxy resin; G, glass fiber weave, CA, carbon/aramid hybrid weave. (From Larsen, T.Ø. et al., *Wear*, 264, 857–868, 2008.)

Both of these factors might increase the deformation, or ploughing, contribution to COF. However, CFs might act as a solid lubricant, which decreases the interfacial shear force due to the partial graphite structure of these fibers. With respect to abrasiveness, CFs are also in some cases observed to roughen the steel counterface but to a lesser extent than the glass fibers. The frictional data do not show any clear variation as a function of load or sliding velocity despite a few exceptions. Thus, it might be concluded that COF is fairly constant and roughly follows Amontons' laws of friction.

From studies on unidirectional fibers, it is known that the fiber orientation with respect to sliding direction can have a large impact on the tribological behavior. Because the wear performance measured for G/EP at the normal and parallel (N–P) fiber orientation is particularly poor, it is chosen to repeat some of the measurements at the parallel and antiparallel (P–AP) fiber orientation in order to determine if changes in the glass fiber orientation have any impact (Figure 3.2). The COF measured for the P–AP direction is independent of the pv condition, whereas the N–P direction shows some variation. Also the specific wear rate $\dot{w}_s$ is roughly the same at the three pv conditions for the P–AP orientation. On the contrary, the N–P direction shows a significant increase in $\dot{w}_s$ when going from a high p–low v condition to a low p–high v condition. As a consequence, $\dot{w}_s$ is a factor of 2.6 higher for the N–P direction compared to the P–AP direction at the condition: $p = 0.25$ MPa, $v = 6.0$ m/s. The latter might be explained by a difference in abrasiveness between the two directions. G/EP is found to be particularly abrasive at this condition, which might account for the high wear rate. The P–AP orientation is, however, found by visual inspection to be less abrasive at the same condition and even forming a grayish transfer film. Additionally, the decrease in COF may also contribute to limit temperature-induced

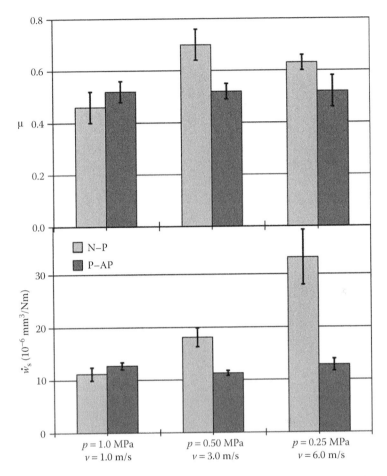

FIGURE 3.2 Comparison of the COF μ and the specific wear rate $\dot{w}_s$ for G/EP when tested with fibers oriented in either the N–P or the P–AP direction. The comparison is made at three different pv conditions. N–P with respect to sliding direction, whereas the fibers are P–AP. (From Larsen, T.Ø. et al., *Wear*, 264, 857–868, 2008.)

deterioration of wear resistance. Relative to the wear rate for the neat epoxy resin, it was found that the P–AP orientation reduced the wear rate by factors of 2.1 and 1.1, respectively, whereas the normal direction increased the wear rate by a factor of 2. According to this, it should be beneficial to orient the weave in a way that eliminates the normal fiber orientation, which agrees well with the data presented here.

As mentioned earlier, development of glass–epoxy composite with fillers has received an increased emphasis for these purposes, and glass–epoxy composite with filler has generated wide interest in automotive and aerospace engineering fields because glass fiber increases the strength and load-bearing capacity in the composite. Addition of solid lubricant tends to reduce the COF and hence reduces the wear loss. In view of the above, several attempts are made to investigate the effect of graphite as

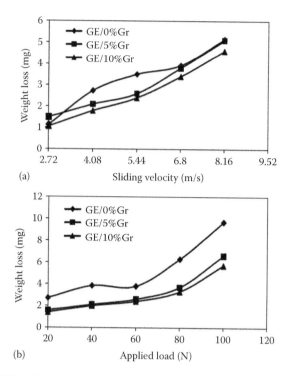

FIGURE 3.3 Effect of (a) the sliding velocity and (b) the applied load on the weight loss of epoxy/glass composites reinforced by graphite. (From Basavarajappa, S. et al., *Mater. Des.*, 30, 2670–2675, 2009.)

a filler material in glass–epoxy composite. Basavarajappa et al. [45] studied the wear characteristics of glass–epoxy composites reinforced by graphite using pin-on-disc apparatus. Figure 3.3 shows the variation of weight loss with respect to the sliding velocity for different volumes of graphite specimens. The sliding velocity is varied from 2.72 to 8.16 m/s, whereas the applied load and sliding distance are kept constant at 60 N and 3000 m, respectively. The experimental data show that the weight loss increases with an increase in the sliding velocity. It is also noticed that 10% graphite-filled composite have less weight loss compared to 5% and 0% graphite specimens. Graphite specimens (5% and 0%) have nearly equal weight loss at lower and higher sliding velocities. However, there is a difference in weight loss at a velocity range of 4–6 m/s. Figure 3.3 shows the variation of the weight loss against the applied load on the filler content of graphite. The applied load is varied, but the sliding velocity and sliding distance are kept constant. Weight loss is dropped from 2.7 to 1.5 mg as graphite content increases from 0 to 10 vol.%. As the applied load increases, the weight loss slightly goes up till 60 N; further increasing the applied load, the weight loss increases sharply. The unfilled glass–epoxy specimen showed the poorest wear performance compared to glass–epoxy filled with graphite specimen. There is a huge difference in weight loss seen between unfilled and graphite-filled glass–epoxy composites, but not much difference is seen between 5% and 10% graphite specimens. At lower load of 40 N, the weight loss of the 5% and 10% graphite specimens was seen to be almost equal; at higher load of 100 N, the 5% graphite specimen had higher weight loss (6.6 mg), whereas the 10% graphite specimen has a lesser weight loss (5.7 mg). The weight loss of the composite with graphite as filler was found to be marginally affected with the change in applied load, whereas in the case of the neat glass–epoxy composite, the variation was substantial. Glass–epoxy composite displays debris formation on the fibers in both transverse and longitudinal directions. Also a few of the broken fibers are seen. At 5% graphite specimen, a complete exposure of fibers is not seen and there is hardly any breakage of fibers.

The effectiveness of replacement of traditional microcomposites with nanocomposites has been verified in many researches that can improve tribological properties. Rong et al. [46] compared the effects of micro-TiO_2 (44 m) and nano-TiO_2 (10 nm) particles on the wear resistance of epoxy. Their results indicate that the TiO_2 nanoparticles can remarkably reduce the wear rate of epoxy, whereas the micron-TiO_2 particles cannot. Similar conclusion was made by Ng et al. [47] in an earlier report. In the work by Yu et al. [48], who studied the tribological behaviors of microscale copper particle- and nanoscale copper particle-filled polyoxymethylene (POM) composites, the wear of micro-Cu/POM is characterized by scuffing and adhesion, whereas that of nano-Cu/POM is by plastic deformation and hence decreased wear loss. Xue and Wang [49] found that nano-SiC is able to lower the wear of polyetheretherketone (PEEK) more greatly than micro-SiC. A thin, uniform, and tenacious transfer film was formed on the counterpart surface in the case of carbon steel ring/nano-SiC filled composite block.

Nanoparticles are effective to improve the tribological properties of polymer composites where nano-SiO_2 particles can also improve the tribological performance of epoxy by changing the wear mechanism from severe abrasive wear to mild sliding wear [50]. As CFs can improve the friction and wear behavior, therefore, these two types of additives are incorporated into the composites; a desirable synergetic effect appears (Figure 3.4) [51]. It is interesting to find that the composites with 4 wt.% nano-SiO_2 and 6 wt.% CF have the lowest specific wear rate, $\dot{w}_s$, and friction coefficient. Compared to the case of unfilled epoxy ($\dot{w}_s = 3.1 \times 10^{-4}$ mm³/Nm, COF = 0.55), the improvement is significant. Although nano-silica is less effective in reducing the wear rate and friction coefficient than short CF, a combination of nano-silica and short CF can get much better outcomes. Moreover, the grafted nano-SiO_2 is superior to the untreated version, reflecting the importance of filler/matrix interfacial bonding. In general, surface hardness is one of the most important factors that govern materials' wear resistance. Harder surface would have higher wear resistance. To understand the synergetic effect revealed in Figure 3.4. CF-reinforced epoxy can be considered as *body* phases dispersed into a *soft* phase [52], which increases hardness and creep resistance of the composites and reduces the wear rate and friction coefficient. However, the addition of nano-SiO_2 decreases the hardness of CF/epoxy composites. Although microhardness of the composites with 4 wt.% nano-SiO_2 and 6 wt.% CF is almost the highest among those containing nano-silica, it is still lower than the composites with 10 wt.% CF. That is, the dependence of microhardness on the weight ratios of the composite

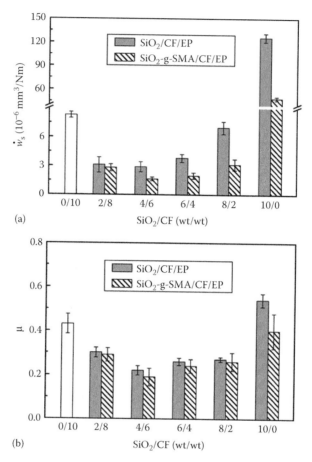

(a)

(b)

FIGURE 3.4 (a) Specific wear rate, $\dot{w}_s$ and (b) friction coefficient of epoxy-based nanocomposites. For reference, the specific wear rate and friction coefficient of unfilled epoxy are listed as follows: 3.1×10^{-4} mm³/Nm and 0.55, respectively. The total filler contents in all the specimens are fixed at 10 wt.%. (From Guo, Q.B. et al., *Wear*, 266, 658–665, 2009.)

components does not exactly agree with the trends in Figure 3.4. Other factors in addition to the enhancement of hardness should account for the improved friction and wear properties of the composites.

To further understand the wear mechanism, the surface of the counterpart steel rings that had rubbed against the composites specimens was also observed by scanning electron microscopy (SEM) (Figure 3.5). Figure 3.5a and b shows that transfer films were discontinuously developed on the steel counterface after sliding on the grafted nano-SiO₂/epoxy and CF/epoxy composites, respectively. For the counterface rubbing against the hybrid composites, uniform transfer films were generated, covering the entire region of interests (Figure 3.5c and d). The formation of transfer film for epoxy composite/steel pairs has long been reported [53]. It certainly favors

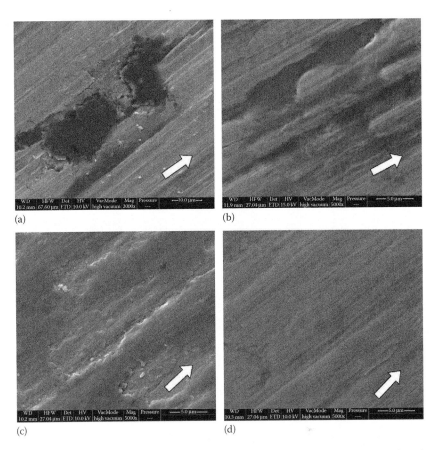

(a)   (b)

(c)   (d)

FIGURE 3.5 SEM micrographs of the steel counterpart surface that had rubbed against (a) SiO_2-g-SMA/EP (SiO_2: 10 wt.%), (b) CF/EP (CF: 10 wt.%), (c) SiO_2/CF/EP (SiO_2/CF: 4/6 [w/w]), and (d) SiO_2-g-SMA/CF/EP (SiO2/CF: 4/6 [w/w]). The arrows indicate the sliding direction. (From Guo, Q.B. et al., *Wear*, 266, 658–665, 2009.)

the reduction in wear rate and friction coefficient. It is believed that for the hybrid composites, CFs take the responsibility of load bearing, whereas the tiny sheetlike wear debris reinforced by nano-SiO_2 act as a polishing agent [54].

To improve the friction and wear behavior of the polymer-based composites, various solid lubricants such as graphite, MoS_2, and PTFE are added. Hexagonal boron nitride (hBN) known as white graphite is perhaps the least explored solid lubricant in spite of its very high thermal stability. It has a lamellar crystalline structure similar to graphite and MoS_2 in which the bond between the molecules within each layer is strong covalent, whereas the binding between layers is almost entirely maintained by means of weak van der Waals forces. Hence, long duration tests (severe operating parameters) were conducted, and the results are shown in Figure 3.6 [55]. In general, while recording the Mu as a function of sliding time, it was observed that it settled down to a very stable value after showing initial fluctuations. This stabilized value

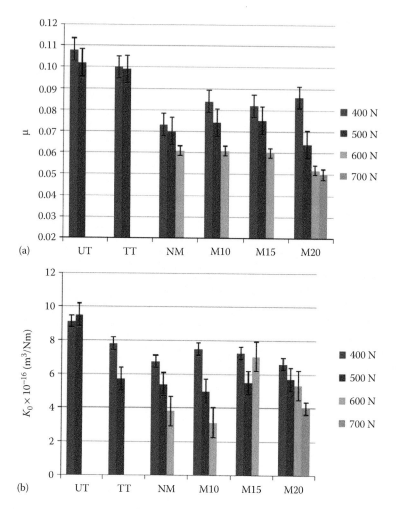

FIGURE 3.6 COF (a) and specific wear rate (b) for selected composites as a function of load under severe operating conditions (speed 1.72 m/s, sliding duration 4 hours, and sliding distance 24.72 km). UT (composite with untreated GrF), TT (composite with treated GrF), NM (composite with treated GrF and top three layers with a mixture of nano- (2%) and micro-hBN (8%); M10 (composite with treated GrF and top three layers with micro-hBN 10%); M15 (composite with treated GrF and top three layers with micro-hBN 15%) and M20 (composite with treated GrF and top three layers with micro-hBN 20%). (From Kadiyala, A.K. and Bijwe, J., *Wear*, 301, 802–809, 2013.)

of COF was considered for plotting the graphs. The salient observations from the studies carried out on the composites are as follows:

The composites showed very low wear rate in the range of $3–10 \times 10^{-16}$ m³/Nm and very low COF (0.05–0.1). The increase in the sliding time keeping other parameters same, led to a slight reduction in COF and an appreciable reduction in the *wear rate* because of more efficient film transfer on the counterface.

Composites UT and TT failed while running under higher load (600 N); UT failed in 60 s and TT within few minutes. Hence, their limiting load was declared as 600 N under a speed of 1.72 m/s. There was again a marginal but positive effect on m due to treatment of fabric. However, the wear rate showed an appreciable reduction (~20%).

With an increase in the load, the m decreased in all the cases substantially as per general trends in the literature. In the case of wear rate, most of the times it reduced with an increase in the load. Once the efficient, uniform, and thin film of graphite (from graphite fiber) is transferred on the disc, the friction is between graphite and graphite. When further improvement in the quality of film is not possible, further improvement in m is also not possible in spite of the increase in sliding time.

Surface treatment of the composites with hBN also led to a significant decrease in COF and wear rate. Among all, NM appears to be the most effective in most of the cases for reducing wear rate. For COF, there was no significant difference because of the size or amount of hBN. It was expected that nano-hBN would show a distinctly different behavior. This could be possibly due to some problem with supplied nanoparticles of hBN as 70 nm size. SEM observation led to a conclusion that it was around 200 nm size. Though the combination of 8% 1.5 mm (1500 nm) along with 2% particles of 200 nm had helped to reduce the wear rate but not COF substantially. If the particles supplied would be of really a nanosize, further enhancement in properties could be possible.

M10 and M15 showed identical COF. However, M20 showed the advantage of the increased amount of hBN by showing the lowest COF (0.05) at 600 N. It showed further reduction in COF at 700 N. Other composites could not be tested under further loads because of nonavailability of samples. Incorporation of hBN on the surface has definitely increased the limiting load of the composite from 500 N to more than 600 N. They did not fail. The wear rates of M10 and NM composites were the lowest, with M10 showing better performance at higher loads, indicating that 10% could be the optimum amount of hBN on the surface for best tribo-properties.

Considering the effect of nanoparticles, MWNT/epoxy nanocomposites with excellent tribological properties have been proposed and studied. Li et al. reported the tribological behaviors of epoxy/CNT composites under dry conditions. It was found that CNTs could dramatically reduce the friction and improve the wear resistance behaviors of the composites, which resulted from the significant reinforcing and self-lubricating effects of CNTs on the polymer matrix. In a research, MWNTs/epoxy nanocomposites were prepared successfully, and the influence of MWNTs reinforcing on the tribological properties of the nanocomposites was investigated. The improved friction and wear mechanisms of the nanocomposites in dry sliding against a plain carbon steel counterpart were also discussed [56]. Figure 3.7a shows the friction coefficient of MWNTs/epoxy nanocomposites as a function of MWNT content for steady-state sliding against the stainless steel ring under dry sliding contact conditions. It is apparent that the friction coefficient of MWNTs/epoxy nanocomposites decreases with increasing MWNT content. The friction coefficient values of nanocomposites sharply decrease when MWNT content is below 1.5 wt.%. As the content of MWNTs in nanocomposites is higher, the friction coefficient becomes lower and keeps a relatively stable value. Figure 3.7b shows the effects of MWNT content on the wear rate of MWNTs/epoxy nanocomposites.

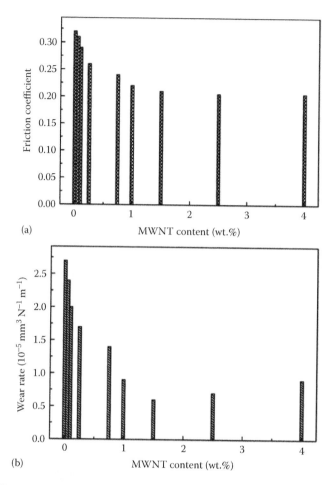

FIGURE 3.7   (a) The friction coefficients and (b) wear rate of MWNTs/EP nanocomposites as a function of MWNT content. (From Dong, B. et al., *Tribol. Lett.*, 20, 251–254, 2005.)

It can be clearly seen that the incorporation of MWNTs significantly decreases the wear rate of epoxy. The wear rate of MWNTs/epoxy nanocomposites decreases sharply from 2.7×10^{-5} to 6.0×10^{-6} mm³/Nm with the concentration of MWNTs from 0 to 1.5 wt.%. It is found that 1.5 wt.% MWNTs/epoxy nanocomposites exhibit the smallest wear rate. When the content of MWNTs in the nanocomposites exceeds 1.5 wt.%, the wear rate of MWNTs/epoxy nanocomposites increases slightly with increasing MWNT content. Similar results were also observed on the microhardness of the nanocomposites.

The SEM images of the worn surfaces of EP and 1.5 wt.% MWNTs/EP nanocomposites are shown in Figure 3.8a and b, respectively. The worn surface of pure EP shows signs of adhesion and abrasive wear (Figure 3.8a). The corresponding surface is very rough, displaying plucked and ploughed marks indicative of adhesive wear and ploughing. This phenomenon corresponds to the relatively poorer

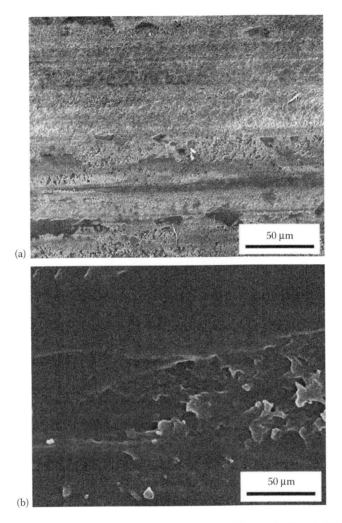

FIGURE 3.8 SEM images of the typical worn surfaces of EP (a) and MWNTs/EP nanocomposites (b). (From Dong, B. et al., *Tribol. Lett.*, 20, 251–254, 2005.)

wear resistance of the pure EP in sliding against the steel. It can be seen that more obvious ploughed furrows appear on the worn surface of the EP block specimen. By contrast, the scuffing and adhesion on the worn surface of 1.5 wt.% MWNTs/EP nanocomposites are considerably reduced (Figure 3.8b). The relatively smooth, uniform, and compact worn surface is in good agreement with the considerably increased wear resistance of the MWNTs/EP nanocomposites. Therefore, it can be deduced that the incorporation of MWNTs contributes to restrain the scuffing and adhesion of the EP matrix in sliding against the steel counterface. As a result, the MWNTs/EP nanocomposites show much better wear resistance than the pure EP. According to some reports [57–61], the prominent friction and wear mechanisms of MWNTs/EP nanocomposites in dry sliding against a plain carbon steel counterpart may be attributed to the following two factors: first, the incorporation of MWNTs in EP helps to considerably increase the mechanical properties of the nanocomposites; hence, MWNTs/EP nanocomposites show much better wear resistance than pure EP. Second, during the course of friction and wear, MWNTs dispersed uniformly in the MWNTs/EP nanocomposites may be released from the nanocomposites and transferred to the interface between the nanocomposites and the steel counterface. Thus, MWNTs may serve as spacers preventing the close touch between the steel counterface and the nanocomposite block, which slows the wear rate and reduces the friction coefficient. Further, the self-lubricate properties of MWNTs also result in the reduction of the wear rate and the friction coefficient. In order to make the mechanism of the tribological performance of MWNTs/EP nanocomposites more clear, further work will be done in our future study.

3.3   POLYTETRAFLUOROETHYLENE

Polymers are extensively used as solid lubricants in dynamic mechanical parts due to their unique properties such as high strength, lightweight and excellent wear, and solvent resistance [62]. PTFE, also named *teflon*, is well known for its extremely low friction coefficient and excellent chemical resistance [63,64]. It is a thermoplastic material with wide-ranging applications in the aerospace, chemical processing, medical, automotive, and electronic industries. PTFE, while being capable of providing low friction in dry sliding conditions, also suffers from an extremely high wear rate that limits its use, whereas the wear rate of PTFE at sufficiently slow sliding speeds is quite low [65]. However, the major drawbacks of PTFE, such as poor wear resistance and severe creep deformation, restrict its wide use in practical applications. Therefore, fibrous fillers (glass fiber, CF, and whisker) [66,67] and spherical nanoparticles [68] are added of PTFE to improve the wear resistance. PTFE has also been demonstrated as an effective filler in other polymers to improve the tribological property of the polymer blends [4,69].

The main objective of PTFE composites is to optimize the wear of the PTFE itself toward low wear while keeping low friction. However, a certain wear is needed to enable the formation of a transfer film on the counter disc or on races in ball bearings. This is needed for low friction (and that will also lower the wear in return). Using pure PTFE would lead to the lowest friction but also to excessive wear, resulting in unacceptable lifetime of space applications. A second aspect is the appearance of the transfer film. There is still an ongoing discussion on the optimum characteristics of such a transfer film [70]. From experience, the hard fillers are necessary to steer this wear process in terms of shape of transfer film and its amount. According to previous studies [8], hard fillers reduce subsurface deformation and *crack propagation*. In addition, the shape of fillers steers the shape of the transfer film, and round fillers are reported to allow a thicker transfer film accompanied by too high wear [71]. Long fillers such as glass fibers are preferred for thin transfer film. However, they may lead to scratching of the counterpart. In order to overcome this risk, a solid lubricant may be added (MoS_2).

Polymeric materials filled with short CF (SCF)/PTFE/graphite are successful tribomaterial formulations [72]. The multiple fillers play synergistic roles in improving the tribological performance of the neat polymer matrix. SCF has been shown to increase the compressive strength and the creep resistance of the polymer matrix. The internal lubricant, that is, graphite and PTFE, reduces the friction coefficient. Moreover, these internal lubricants contribute to the formation of a homogeneous transfer film on the counter surface. This can reduce the friction coefficient and wear rate by avoiding the direct contact between the friction pairs.

Nanoscale fillers have been thought to reduce wear by a number of mechanisms, including preventing frictional destruction of PTFE banded structure [73], enhancing transfer film-counter surface adhesion [74], or inducing changes to the PTFE matrix structure that are more wear resistant [11,75,76]. Chen et al. [59] studied the friction and wear properties of PTFE/CNT composites using a ring-on-block arrangement under dry condition. It was found that CNTs significantly increased the wear resistance of PTFE composites and decreased their friction coefficient, which was attributed to the super strong mechanical properties and the very high aspect ratio of CNTs. Moreover, graphene is intriguing from the point of view of wear suppression because it has in-plane dimensions on the order of several microns coupled with nanometer-scale sheet thickness. The microscale dimensions of the graphene sheets might enable them to effectively interfere with the debris generation processes in polymers, whereas the nanometer-scale thickness, low density, and planar geometry of graphene generate a huge interfacial contact area with a very large number density of graphene sheets in the polymer matrix.

A research study reported the ability of graphene platelets to drastically reduce wear rates in PTFE to as low as $\sim 10^{-7}$ mm^3/Nm, which is 4 orders of magnitude lower than in baseline PTFE. The excellent wear suppression performance of graphene additives coupled with their potentially low production cost (top–down synthesis by exfoliation of graphite) makes this technology very promising for industrial-scale applications [65].

The wear volume-sliding distance records for unfilled PTFE and PTFE filled with varying amounts of graphene are shown in Figure 3.9a. Unfilled PTFE wears rapidly, accumulating ~ 33 mm^3 of the worn volume after a mere ~ 1.5 km of sliding. It is clear that the 0.02, 0.05, and 0.12 wt.% graphene composites also show rapid wear behavior, with their data points crowding near the vertical axis similar to those of unfilled PTFE. The effectiveness of the graphene as a potential wear suppressant is first seen in the 0.32 wt.% composite, which wears more gradually than unfilled PTFE, needing ~ 20 km of sliding before accumulating as much wear volume as unfilled PTFE did after ~ 1.5 km. Further improvement in wear resistance is witnessed for the 0.8 and 2 wt.% composites, with the former accumulating only ~ 12 mm^3 in wear volume after 26 km of sliding and the latter losing ~ 4 mm^3 after ~ 51 km of sliding. This trend of increase in wear resistance with increasing filler contents continued for the 5 and 10 wt.% composites with their data points crowding near the horizontal axis as a consequence of their extremely slow wear rates. The responses of these highly wear-resistant 2, 5, and 10 wt.% composites are more clearly visible in Figure 3.9b, which are obtained by expanding the scale of the wear volume axis from Figure 3.9a. It is clear from Figure 3.9b that even the low wear 2, 5, and 10 wt.% composites initially show a transient higher wear rate behavior, during which they accumulate

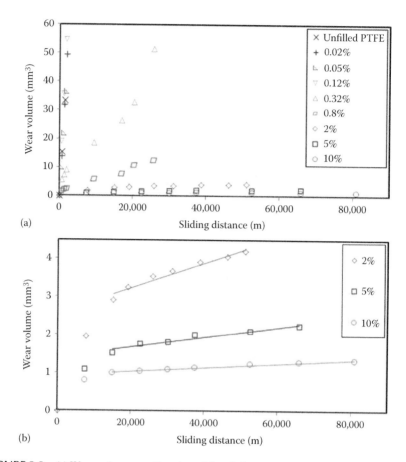

(a)

(b)

FIGURE 3.9 (a) Wear volume as a function of the sliding distance for unfilled PTFE and for PTFE with varying contents (wt.%) of graphene platelet filler. (b) The wear volume-sliding distance records for the slow wearing 2%, 5%, and 10% composites. The steady-state behavior for each composite is indicated by a trendline, the slope of which was used to compute the steady-state wear rate for the composite. The uncertainty in the wear volume measurements was ±0.05 mm^3. (From Kandanur, S.S. et al., *Carbon*, 50, 3178–3183, 2012.)

a few mm^3 of wear volume, before they transition to a lower steady-state wear behavior that is indicated with the help of trendlines. The transition from transient behavior to steady-state behavior for the composites in Figure 3.9b seems to occur within the first 7–15 km of sliding.

Wear tracks on the counter surfaces of the high wear rate unfilled PTFE and the extremely low wear rate 10 wt.% graphene/PTFE composite. Large plate like wear debris, hundreds of micrometers in in-plane dimensions, is seen throughout over the counter surface of the rapidly wearing unfilled PTFE (Figure 3.10a). Noticeably, smaller wear debris, generally much less than 100 Mu in dimensions, is seen on the counter surface of the low wear rate 10 wt.% graphene–PTFE composite along the

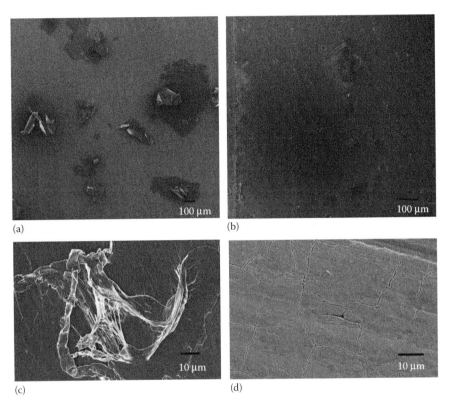

FIGURE 3.10 (a) SEM micrograph of the counter surface of unfilled PTFE showing wear debris hundreds of micrometers in in-plane dimensions. (b) The counter surface of the low wear rate 10% graphene platelet–PTFE showing wear debris that is finer compared to that generated by the rapidly wearing unfilled PTFE. (c) Wear surface of unfilled PTFE showing large platelike debris on the surface. (d) Corresponding wear surface of the 10% graphene platelet–PTFE composite displaying wear-resistant *mudflat* features. (From Kandanur, S.S. et al., *Carbon*, 50, 3178–3183, 2012.)

edges of the wear track (running along the left side of Figure 3.10b). Figure 3.10c and d presents the wear surfaces of the unfilled and graphene-filled PTFE. A top the otherwise smooth wear surface of the unfilled PTFE, the large plate-like debris are again seen (Figure 3.10c), either in the process of detachment or as back-transferred debris cycling in attachment to the counter surface then returning to the polymer surface until eventual ejection from the contact. The wear surface of the graphene-filled PTFE displayed a *mudflat cracking* appearance, as also reported by Burris et al. [22] for comparably wear-resistant PTFE when filled with 80 nm alpha-phase alumina particles. These mudflat surface regions are of approximately 10 μm scale and, as also reported for the wear-resistant alumina-filled PTFE nanocomposites [22], these regions appear to be secured into place by PTFE fibrils, which span the shallow cracks separating them.

3.4 POLYETHERETHERKETONE

PEEK displays the combination of good tribological performance and high strength, and therefore, it is being widely used as a tribomaterial [77]. The tribological performance of a material is affected by the processes that occur in the surface layers formed during sliding. The tribological characteristics, that is, friction coefficient and wear rate, strongly reflect the surface behavior of the material rather than their bulk properties. Therefore, the tribological performance of SCF-filled PEEK composites cannot be correlated directly to their bulk properties such as compression strength and impact strength [78]. For the same reason, fiber orientation can exert a different effect on the tribological performance compared with its effect on the bulk properties of fiber-reinforced polymers, for example, tensile and compression properties. The stresses, depending on fiber orientation, in the surface layer of unidirectional fiber-reinforced polymer composite slid over rough surfaces play an important role in fiber failures. Cirino et al. [79] investigated the effect of fiber orientation on the abrasive wear behavior of 55 vol.% unidirectional continuous CF (CCF)-reinforced PEEK by sliding the composite on abrasive papers. Their results indicated that the wear resistance of the composite was best when the fiber orientation was normal (N orientation) to the sliding direction. The composite displayed the highest wear rate when the CCF orientation was antiparallel (AP orientation) to the sliding direction. Friedrich [80] investigated the effect of fiber orientation on the sliding behavior of 55 vol.% CCF-filled PEEK against smooth steel. The results showed that the optimum wear resistance occurred when the fiber orientation was parallel (P orientation) to the sliding direction and that the composite exhibited the worst wear resistance when the fiber orientation was normal to the sliding direction (Figure 3.11). Voss and Friedrich [72] investigated the effect of fiber orientation on the wear behavior of 30 wt.% short glass fiber (SGF; ~18 vol.%)-filled PEEK composites. It was found that when slid against smooth steel, the composite exhibited the best wear resistance with a sliding direction normal to the fiber orientation. However, similar results with no systematic difference in the value of wear rates were obtained for P and AP orientations. Summarizing the earlier results, it is uncertain which fiber orientation contributes the best to the wear

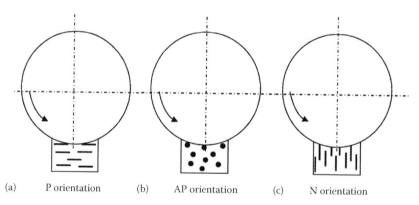

(a)   P orientation   (b)   AP orientation   (c)   N orientation

FIGURE 3.11   (a–c) Designations of fiber orientations. (From Zhang, G. et al., *Wear*, 268, 893–899, 2010.)

resistance of fiber-reinforced polymers. The fiber type (e.g., GF or CF), the aspect ratio of fiber (e.g., SCF or CCF), the fiber fraction, and the wear conditions (sliding wear or abrasive wear in which the counter surface roughness is much larger than the former) can all be the influential factors determining the effect of fiber orientation. With regard to SCF-reinforced polymers, the matrix wear and the fiber wear thinning, cracking, and debonding are the important mechanisms determining the tribological performance of the composite [20,81,82]. Especially at a high contact pressure, the cracking and debonding of fibers are the important factors determining the wear rate of SCF-reinforced polymers. Taking into account the fact that fibers support most loads during the sliding process, the CF fraction, the fiber/matrix adhesion, and the nominal pressure can influence the loading state and the failure of the fiber. These factors can also influence the effect of fiber orientation on the tribological performance.

SCF/PTFE/graphite (10 vol.% for each filler)-filled PEEK is widely used as a special tribomaterial grade. However, the effect of SCF orientation, especially for the PEEK system with a small SCF fraction, has been rarely studied. The objective of this study is to investigate the effects of fiber orientation and nominal pressure on the tribological behavior of SCF/PTFE/graphite (10 vol.% for each filler)-filled PEEK [83]. Figure 3.12a shows the mean friction coefficients as functions of apparent pressure and fiber orientation. In the P and N orientations, the friction coefficients increase with increasing nominal pressure. At 1 MPa, the friction coefficient in the N orientation is lower than that in the P orientation but similar to that in the AP orientation. At nominal pressures higher than 2 MPa, the friction coefficients in the P and N orientations are similar. At nominal pressures higher than 3 MPa, the friction coefficients in the AP orientation are remarkably lower than those in the P and N orientations. This trend is more pronounced under high pressures. Figure 3.12b shows the wear rate of the composite as functions of pressure and fiber orientation. In the P orientation, similar to the friction coefficients, the increase in apparent pressure from 1 to 4 MPa leads to higher wear rates. When the pressure is increased from 4 to 5 MPa, the wear rate is slightly decreased. In the AP orientation, the increase in pressure from 1 to 2 MPa results in an increase in the wear rate. However, further increasing the pressure from 2 to 5 MPa does not significantly influence the wear rate. In the N orientation, the increase in pressure from 1 to 5 MPa monotonically increases the wear rate. Under low pressures ranging from 1 to 2 MPa, the wear rates in the three orientations are similar. Under high pressures, however, the wear rates in the AP orientation are significantly lower than those in the P and N orientations. From the earlier results, it is clear that the fiber orientation exerts an important influence on the tribological performance of the composite and that the effect of fiber orientation shows a strong dependence on the nominal pressure. In this study, especially at high nominal pressures, the composite displays the best tribological performance in the AP orientation. This tendency is different from 55 vol.% CCF-filled PEEK [80] and 30 wt.% SGF-filled PEEK. This difference can be attributed to several aspects such as different fiber length, fiber fraction, fiber/matrix adhesion, and fiber type.

The SCF supports most of the load and is more wear resistant than the matrix. Therefore, wear debris tends to accumulate near the SCF. In this case, fiber thinning (fiber wear) can be a dominant factor determining the wear rate [81]. P orientation compared with the AP orientation, much less wear debris is observed near the SCF

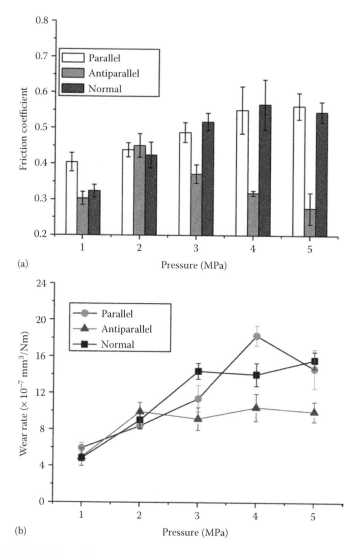

(a)

(b)

FIGURE 3.12   (a) Mean friction coefficients in the three orientations as a function of apparent pressure. (b) Mean specific wear rates in the three orientations as a function of apparent pressure. (From Zhang, G. et al., *Wear*, 268, 893–899, 2010.)

in the P orientation. Moreover, in the P orientation, as indicated by dashed arrows on the worn surface, breakup and removal of CF are noticeable on the worn surface. The N orientation, similar to the AP orientation, wear debris tends to accumulate near the SCF and fiber failure is hardly observed. It is generally accepted that fiber failures lead to a higher relative wear rate. From Figure 3.12, it is interesting to note that regardless of the fiber failures occurring in the P orientation, the wear rates in the AP and N orientations are similar or even slightly higher (<2 MPa) than in the P orientation. This might infer that under low pressures, besides the SCF failures, fiber thinning is also an important factor influencing the effect of fiber orientation. CF has a layered structure consisting of carbon layers in which the carbon atoms are arranged in a hexagonal unit cell [84]. Different carbon layers are linked to each other by weak van der Waals bonds. In CF, carbon layers are preferentially parallel to the fiber axis. As a result, CF exhibits highly anisotropic properties: it displays much stronger mechanical properties, for example, tensile and compressive strengths, along the fiber axis than perpendicular to the axis. Although no direct proof is found in literature studies concerning the tribological anisotropy of an individual fiber (no matrix), it can be assumed that much stronger mechanical properties along the fiber axis result in a higher wear resistance in this direction. If this is the case, the SCF wear is slower along the fiber axis than perpendicular or normal to the fiber axis. In the tribological system of the composite, under a low pressure where the SCF failure is not severe yet, the lower SCF wear rate in the P orientation can benefit the wear resistance. Therefore, the SCF wear and SCF failures are the competing factors to the effect of fiber orientation.

3.5 PHENOLIC

Fabric-reinforced polymer composites have recently generated immense commercial and academic interest due to their wide applicability in the fields of aircraft, aviation, high-speed railway, automobile, and so on [85–87]. They exhibited enhanced mechanical strength in both longitudinal and transverse directions of the fabric and possessed the ability to conform to curve surfaces without wrinkling, compared with other polymer composites [88].

The fabric self-lubricating liner is a kind of woven polymer composite, which is widely used in numerous industrial fields, such as self-lubricating spherical plain bearings, journal bearings, and aerospace [89–91]. As a solid lubrication material, the fabric self-lubricating liner is characterized by low friction, high mechanical strength and impact resistance, good designability, and cost-effectiveness. Nanotechnology offers a novel route to fabricate new materials with excellent mechanical, thermal, electrodynamic, and tribological properties [92–94]. This has drawn a great attention for over a decade. The combination of well-dispersed nanofillers/nanomaterials and polymers gives birth to a new hybrid material–polymer nanocomposite, which has provided increasing applications in different fields [95–97].

Accordingly, silicate mineral becomes a more competitive nanofiller candidate because it has been successfully incorporated into continuous polymer matrices as reinforcements/fillers for a very long time [98,99]. Taking advantage of the broadened interlayer spacing of well-dispersed silicate minerals, intercalated structure occurs when polymer chains are penetrated into interlayer areas to form polymer/layered silicate nanocomposites. The natural layered structure of silicate makes it possess good solid lubrication [100] with the well-studied interaction mechanism [101].

Organic montmorillonite (OMMT)/phenolic (PF) nanocomposites, which worked as hybrid matrices to be combined with the woven fabric for fabricating self-lubricating liner. Several tribological tests were performed to evaluate

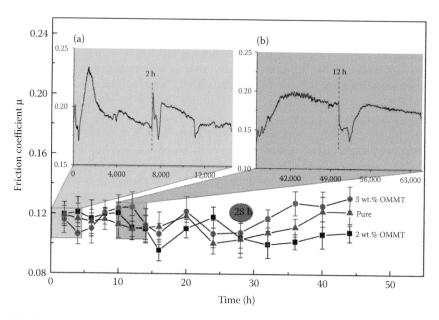

FIGURE 3.13 Average friction coefficient of different fabric self-lubricating liners under long-term friction (a) 0–4 h and (b) 10–14 h. (From Fan, B. et al., *Tribol. Lett.*, 57, 22, 2015.)

their tribological performances. An investigation tried to extend the service life of products containing fabric self-lubricating liner by means of improving the tribological properties of liner composites [102]. The average friction coefficient variations of different fabric self-lubricating liners and the friction coefficient curves during 0–4 hours (Figure 3.13a) and 10–14 hours (Figure 3.13b) for the fabric self-lubricating liner with the OMMT content of 2 wt.% are shown in Figure 3.13. Friction coefficients of all liners are basically at the same level at the early friction stage. As the friction time is more than 28 hours, the average friction coefficients of the liners are apparently kept separate: a liner with the OMMT content of 2 wt.% shows the lowest friction coefficient, whereas a liner with the OMMT content of 5 wt.% exhibits the highest friction coefficient. Figure 3.14 shows the wear losses of three kinds of fabric self-lubricating liners. As observed in the figure, the wear process of both pure liner and liner with 5 wt.% OMMT was basically split into two stages through the experimental friction time: violent wear with rapid wear (0–8 hours) and mild wear with slight and steady wear (8–44 hours). However, wear loss curves of liner with 2 wt.% OMMT seemed to present a monotone-increasing trend through the experimental time without any transition. At early wear stage (0–8 hours), wear losses of both pure liner and liner with 5 wt.% OMMT suffered from a rapid consumption, but the wear loss of the liner with 2 wt.% OMMT was lower than that of the others. Hereafter, the wear loss level was distinct in subsequent wear. The liner with 2 wt.% OMMT shows the best wear resistance (lowest wear loss), followed by the liner with 5 wt.% OMMT and pure liner. As the test finished after 44 hours, wear losses of the pure liner and liners with 2 and 5 wt.% OMMT were 0.233, 0.132, and 0.170 mm, respectively.

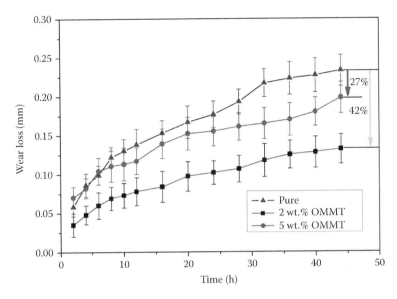

FIGURE 3.14 Wear losses of different fabric self-lubricating liners under long-term friction. (From Fan, B. et al., *Tribol. Lett.*, 57, 22, 2015.)

Compared with the pure liner, wear losses of the liners with 2 and 5 wt.% OMMT were decreased by 42% and 27%, respectively. Overall, the appropriate addition of OMMT improves the wear resistance of fabric self-lubricating liner, particularly at the OMMT content of 2 wt.%.

Figure 3.15a shows the wear losses of different liners at 2 hours of the friction time. It is clear that the liner with 2 wt.% OMMT exhibits the best wear resistance, and both liners with OMMT suffer from less wear loss than the pure liner. Moreover, Figure 3.15b compares the wear losses of liners with 2 and 5 wt.% OMMT at different friction times. Obvious superiority of the liner with 2 wt.% OMMT is observed, and this superiority is enlarging with friction time. In addition, the lowered shear deformation of OMMT/polymer composites decreases friction and/or wear of the liner. However, at a higher content of OMMT, the interlayer spacing of OMMT decreased and the mobility of polymer chains was reduced, which hindered the penetration of polymer chains into the interlayer areas of silicate particles and thus deteriorated the uniform particle dispersion in polymer matrices. Consequently, the particle agglomeration was more pronounced [103]. Particle agglomeration increases the surface roughness and breaks the uniformity of the transfer film and the tribosurface, which in turn accelerates both friction and wear of the liner.

Nomex fiber, a kind of aramid fiber, possessed the properties of high strength, high flame resistance, stable chemical structure, and good resistance to chemicals and abrasion, and these outstanding properties made Nomex fabric a good candidate for developing fabric-reinforced polymer composite [104–107]. Consequently, Nomex fabric or Nomex/phenolic composites were developed and employed in both military and industrial fields as garments, liners, and structure materials [108–111],

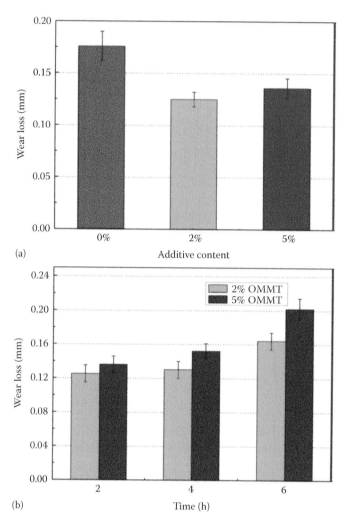

FIGURE 3.15 (a) Wear losses of different liners under high-velocity/light-load friction and (b) wear losses of the liner with 2 and 5 wt.% OMMT. (From Fan, B. et al., *Tribol. Lett.*, 57, 22, 2015.)

where excellent anti-wear and load-bearing properties were required. However, the inertness of Nomex fiber surface hinders the tight adhesion between fabric and adhesive resin, and thus influences the wear property of Nomex fabric-reinforced polymer composites. Thus, seeking a way to improve the wear properties of the Nomex fabric-reinforced polymer composites becomes an urgent demand. Filler reinforcing is a universal way of improving the tribological property of the fabric-reinforced polymer composites. The addition of fillers into the matrix can usually enhance load-withstanding capability, reduce COF, and improve wear resistance and thermal properties [112,113]. As the building block of graphite, graphene displays laminated structure, low shear strength, and thermal stability, which enable it to be a good choice for solid lubricant [114–118]. Recently, it has been demonstrated that graphene-filled polymer composites showed enhanced tribological property with the friction coefficient and wear rate reduced significantly, compared to the original polymer composites [65,119,120].

Self-lubricating composites were prepared by graphene and polystyrene-functionalized graphene (PS-graphene) as fillers to improve the tribological properties of Nomex fabric/phenolic composites. Wear tests showed that the tribological properties of both graphene and PS-graphene-filled fabric/phenolic composites were optimized, compared with unfilled and graphite-filled fabric composite. The effect of filler content, applied load, and sliding speed on the tribological properties of the Nomex fabric/phenolic composites was investigated. Based on the characterizations, the probable reasons of the reinforcement were discussed [111].

The friction coefficient, wear rate, and bonding strength of the unfilled and different filler content-filled Nomex fabric composite are shown in Figure 3.16. It was indicated that the filler content has a great influence on the friction and wear behaviors of fabric composites. All lubricant-filled Nomex fabric composites displayed much lower friction coefficient than the unfilled Nomex fabric composites. The wear rate of the Nomex fabric composite was reduced when 1 wt.% lubricants were filled, and when the mass fraction of lubricants was increased to 2 wt.%, the filled Nomex fabric composite exhibited a lowest wear rate. It was found that 2 wt.% PS-graphene-filled Nomex fabric/phenolic composite showed lower friction coefficient and better anti-wear property than 2 wt.% graphene-filled and 2 wt.% graphite Nomex fabric/phenolic composite. However, further increasing the filler content, the wear rate of the filled Nomex fabric composite increased, which resulted from the weakened bonding strength of the composite.

As shown in Figure 3.17a and d, the surface of the unfilled Nomex fabric composite is smooth and without flakes attached. For graphene-filled Nomex fabric

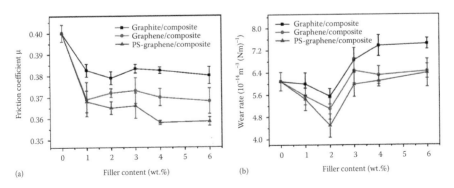

FIGURE 3.16 The value of friction coefficient (a) and wear rate (b) for different filler content-filled Nomex fabric composites. The load and sliding speed in the tests were 0.28 m/s and 110 N, respectively. (From Ren, G. et al., *Compos. Part A: Appl. Sci. Manufact.*, 49, 157–164, 2013.)

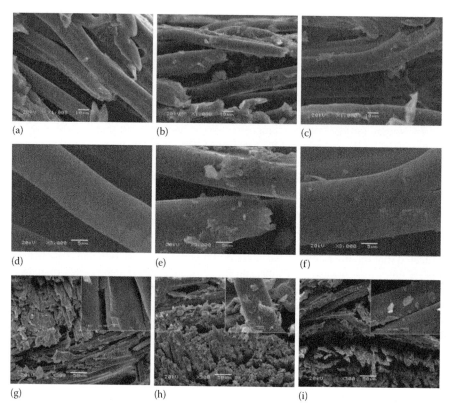

FIGURE 3.17 SEM images of (a) unfilled Nomex fabric composite, (b) graphene-filled Nomex fabric composite, and (c) PS-graphene-filled Nomex fabric composite. (d–f) Magnified images of (a), (b), and (c), respectively; (g–i) fracture surfaces of unfilled Nomex fabric composite, graphene-filled Nomex fabric composite, and PS-graphene-filled Nomex fabric composite, respectively. (From Ren, G. et al., *Compos. Part A: Appl. Sci. Manufact.*, 49, 157–164, 2013.)

composites, graphene nanosheets attached onto the surface of Nomex fiber and were dispersed in phenolic resin (Figure 3.17b), according to the published study [121,122]. However, it can be clearly seen that graphene nanosheets aggregated on the fiber surface (Figure 3.17e), which may hinder the compact adhesion between Nomex fabric and phenolic resin. As for the PS-graphene-filled Nomex fabric composite, aggregations of graphene nanosheets became fewer and smaller (Figure 3.17c and f), and the rest graphene nanosheets were dispersed evenly on the fiber surface and in the phenolic resin. Moreover, the varied fracture surfaces between graphene-filled and PS-graphene-filled Nomex fabric composite also confirmed that PS-graphene was dispersed much more uniformly than graphene in the Nomex fabric composite (Figure 3.17g–i). In addition, the filler/matrix bonding strength of PS-graphene-filled Nomex fabric composite was proved to be better than that of graphene-filled Nomex fabric composite owing to the stronger mechanical interlocking and covalent bonding at the interface [123–125]. The improved dispersibility and the enhanced

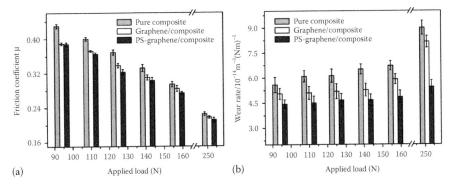

FIGURE 3.18 Friction coefficient (a) and wear rate (b) of unfilled, graphene-filled, and PS-graphene-filled Nomex fabric composite as a function of the applied load. The sliding speed in the tests was 0.28 m/s. (From Ren, G. et al., *Compos. Part A: Appl. Sci. Manufact.*, 49, 157–164, 2013.)

interfacial bonding strength were all beneficial for reducing the friction coefficient and improving the anti-wear property of Nomex fabric composite.

Then, the effect of applied loads on the friction and wear behaviors of unfilled, graphene-filled, and PS-graphene-filled fabric composite was investigated, and the results are shown in Figure 3.18. For all the three composites, the friction coefficient decreased with increasing applied load (Figure 3.18a), whereas the wear rate increased with increasing applied loads (Figure 3.18b). However, the wear rate for the PS-graphene-filled fabric composite increased slightly with increasing applied loads. The friction coefficient and wear rate of composite and the temperature of the worn surfaces followed the order of PS-graphene-filled composite < graphene-filled composite < unfilled composite under every applied load. Thus, the PS-graphene-filled composite possessed enhanced anti-wear property and load-carrying capacity especially under high load.

The influence of the sliding speed on the tribological properties of the three Nomex fabric composites was also investigated, at a fixed load of 94.1 N, and the result is shown in Figure 3.19. It can be seen that the friction coefficient for all the three composites differed slightly with a sliding speed in the range of 0.28–0.73 m/s. The wear rate of the unfilled composite increased obviously with increasing sliding speed. For filled composites especially for PS-graphene-filled composite, the wear rate differed slightly even with a sliding speed of 0.73 m/s. Among all the sliding speeds investigated, the friction coefficient and wear rate of composite and the temperature of the worn surfaces followed the order of PS-graphene-filled composite < graphene-filled composite < unfilled composite. It is believed that with the increase of the sliding speed, transfer film can form more easily on the frictional surfaces, which will improve the lubrication condition at the rubbing surfaces and thus lead to the decrease of the friction coefficient. Adversely, the wear rate increased with the increase of the sliding speed. It is likely that there will be over abundant friction heat after sliding for a longer journey at a higher speed which may cause the reduction of mechanical strength and the load-supporting capacity of composite.

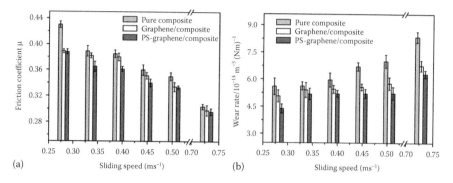

FIGURE 3.19 The effect of the sliding speed on the friction coefficient (a) and wear rate (b) for unfilled, graphene-filled, and PS-graphene-filled Nomex fabric composite. (From Ren, G. et al., *Compos. Part A: Appl. Sci. Manufact.*, 49, 157–164, 2013.)

3.6 POLYIMIDE

Polyimide (PI) is a leading engineering polymer because of its outstanding combination of performance and ease of synthesis [126–128]. It has very extraordinary comprehensive performance, such as excellent mechanical properties, high dielectric properties, prominent thermal endurance, acid-resistant and alkali-resistant properties, and good friction lubrication characteristics under low or high temperature [129,130], which have found wide applications in aerospace, automobile, and microelectronics industry. In some cases, it has replaced traditional steels, which is of great significance in view of the decreasing availability of steels [131]. However, the intrinsic large friction coefficient and high wear rate of pure PI limit its use in dynamic motion systems and a bearing material [132]. Tremendous efforts have been devoted to reduce the friction coefficient and wear rate of PI by incorporating fibers [133], nanometer particles [134,135], solid lubricant [136], and so forth.

Introducing various fibers or particles into PI matrix has been verified to be an effective way to improve the tribological properties of PI. Using polymer blending process also can achieve the target. On account of good lubrication of PTFE, the transfer film (hybrid PI and PTFE) formed on the counterface has a low coefficient of friction and high wear resistance, and the polymer alloy of PI/PTFE composite shows improved tribological properties compared with pure PI and pure PTFE, respectively [137,138]. Some fibers such as carbon fiber, glass fiber, and carbon nanotube usually are incorporated into polymers. Because of tangled and load-carrying fibers, the mechanical properties of polymer-based composite filled with fibers are improved. Meanwhile, even when fibers break, they can still maintain good carrying capacity, which results in high wear resistance of polymer/fiber composites [139–141].

Almost all types of nanoparticles can be used as reinforcement phase for polymer-based composites, such as Al_2O_3, ZrO_2, ZnO, TiO_2, and Cu [142–145]. The rigid nanoparticles incorporated into polymer matrixes can act as support loads, and also can form micro-bearings on the counterface, which leads to a low coefficient of friction. Moreover, the self-lubricating transfer film generated on the counterface separates polymer-based composites from couple pairs, which is beneficial to reduce the wear of composites [146,147]. However, not all nanoparticles will lead to high tribological properties for polymer-based composites. Some nanoparticles added into polymer matrix, to a certain extent, also deteriorate the tribological properties of polymer-based composites [148,149]. However, the addition of some nanoscale solid self-lubricating materials for improving the tribological properties of polymer composites is indeed very significant, which is mainly due to the self-lubricating properties of added materials [150,151].

Zinc oxide (ZnO), with outstanding mechanical properties [152], has been widely used as a reinforcing filler in various polymers. Li et al. [73] reported that the wear volume loss of 15 vol.% ZnO/PTFE is only 1% compared to pure PTFE. Chang et al. [153] found that 10 wt.% ZnO nanoparticles filled in ultra-high molecular weight polyethylenes (UHMWPE) reach the optimal tribological and mechanical properties of the composites with 30% and 190% enhancement in wear resistance and compressive strength, respectively, compared to the pure UHMWPE. To the best of our knowledge, the effect of ZnO nanoparticles on the tribological and mechanical properties of PI-based nanocomposites has rarely been studied. In this investigation, PTFE/PI blend polymer was reinforced by different loadings of ZnO nanoparticles. The optimal loading was explored in association with greatest tribological and mechanical properties. The microstructures of the worn surface, transfer film, and impact-fractured surface were also examined to understand the reinforcing effect of ZnO in the nanocomposites [154]. Figure 3.20 shows the wear volume loss and friction coefficient of ZnO/PTFE/PI nanocomposites as a function of ZnO. With increasing ZnO loading, both wear volume loss and friction coefficient decrease and reach the minimum value at 3 wt.% and continuously go up afterward. Specifically,

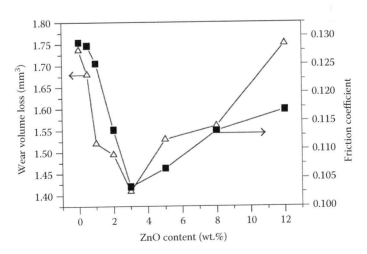

FIGURE 3.20 Effect of ZnO loading on wear volume loss and friction coefficient. Load: 100 N; sliding speed: 1.4 m/s. (From Mu, L. et al., *J. Nano.*, 16, 373, 2015.)

the wear volume loss of the nanocomposites reinforced with 3 wt.% ZnO is 20% less than that of the PTFE/PI blend polymer. With further increasing ZnO loading, polymer chains are not sufficient to cover the extremely large surface area exposed by ZnO nanoparticles, and thus, nanoparticles tend to agglomerate and negatively affect the interfacial bonding with polymer matrix. The weak interfacial bonding becomes weak joint, which can be easily peeled off, and thus, worse tribological property is observed.

The effects of the sliding speed on the wear volume loss and friction coefficient of various specimens at a load of 200 N are shown in Figure 3.21. It can be seen that

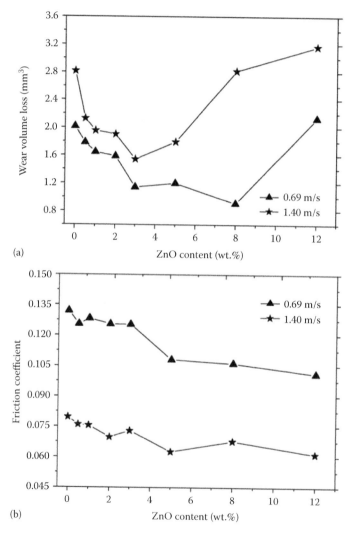

FIGURE 3.21   (a) Wear volume loss and (b) friction coefficient with ZnO loading at 1.4 and 0.69 m/s. Load: 200 N. (From Mu, L. et al., *J. Nano.*, 16, 373, 2015.)

the variation of both wear volume loss and friction coefficient experiences the same trend under sliding speeds of 0.69 and 1.4 m/s. The wear volume loss of ZnO/PTFE/PI nanocomposites decreases to the minimum value and then increases afterward. The lowest wear loss is achieved with 3 wt.% ZnO loading at a higher sliding speed of 1.4 m/s, whereas larger ZnO loading of 8 wt.% is required to reach the minimum value at a relatively lower speed of 0.69 m/s. The ZnO nanoparticles serve as rolling balls between the friction interfaces, which would definitely reduce the interfacial friction and improve the tribological properties [155]. At higher sliding speed, larger shear force and friction energy facilitate the *pulling off* of ZnO nanoparticles from polymer matrix and accumulation at the interface. Therefore, optimal amount of *rolling* ZnO nanoparticles would be accumulated at the interface from the nanocomposites with relatively lower ZnO loadings [156].

Figure 3.22 displays the optical micrographs of transfer films formed on the counterpart steel ring after friction test. The transfer film of PTFE/PI appears to be rough and discontinuous (Figure 3.22a), which could be easily scaled off from the wear track, and negatively affects the wear resistance during sliding. Apparently, the transfer film for 3 wt.% ZnO is continuous, uniform, and smooth, which is helpful to maintain a stable friction, and thus, the best tribological properties are obtained

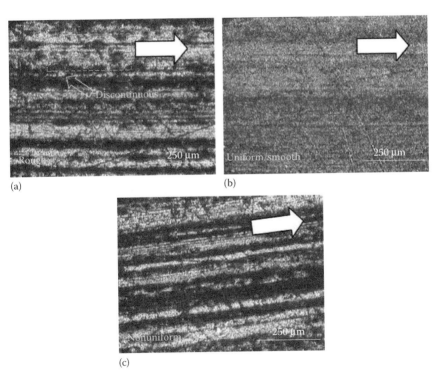

(a)   (b)

(c)

FIGURE 3.22 Optical micrographs of transfer films (200×, 100 N, 1.4 m/s): (a) PTFE/PI, (b) 3 wt.% ZnO/PTFE/PI, and (c) 12.wt.% ZnO/PTFE/PI. Arrow indicates the sliding direction. (From Mu, L. et al., *J. Nano.*, 16, 373, 2015.)

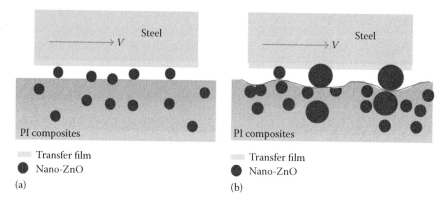

FIGURE 3.23 Schematic illustration of the role of ZnO at (a) low and (b) high volume fraction. (From Mu, L. et al., *J. Nano.*, 16, 373, 2015.)

(Figure 3.22b). When ZnO loading increases to 12 wt.%, the transfer film becomes nonuniform, which reveals that an excess amount of ZnO would hinder the formation of a smooth transfer film (Figure 3.22c). As a result, the wear volume loss was increased considerably at higher filler content. These results reveal that a suitable loading of ZnO in the PTFE/PI polymer blend will favor the formation of smooth and continuous transfer films and contribute to the enhanced tribological properties.

Figure 3.23 illustrates the role of ZnO nanoparticles during friction at different loadings. *Rolling ball* effect dominates the interfacial friction at relatively lower ZnO loading due to the good dispersion and particle–polymer interaction (Figure 3.23a). Excess amount of ZnO in the nanocomposites leads to severe agglomeration, which affects the integrity of the nanocomposites in bulk and damages the transfer films at the friction interface (Figure 3.23b). Therefore, peeled debris has been observed on the worn surface (Figure 3.5c) and also rough and noncontinuous transfer film on counterpart ring (Figure 3.22c).

Carbon spheres, which are one of the allotropic substances of carbon, are often used as a template preparation of hollow spherical materials owing to the nature of the surface easily changeable. They possess small density, large specific surface area and chemical stability, and other merits. Therefore, they also can be applied in the area of electrochemistry as an electrode material because of good conductivity and stability [157,158]. Although carbon spheres have been widely studied in the many areas [159–163], the tribological studies of carbon spheres are scarce. Recently, neat C60 had already been shown to have a fairly low friction coefficient at least under certain conditions [164]. Pozdnyakov et al. [165] investigated the sliding friction and wear characteristics of pyrimidine-containing PI coatings and PI–C60 composite. The result showed that the introduction of C60 could improve further the wear characteristics of the coatings down to specific wear rates below $2\sim10^{-7}$ mm^3/Nm and suggest that the role of the C60 molecules in reducing the wear of composite coatings is due to the interactions between C60 and the PI. Due to C60 which is a kind of carbon spheres, the tribological behavior of polymers also might be improved through the introduction of carbon spheres into a mechanically and thermally stable polymer, such as PI.

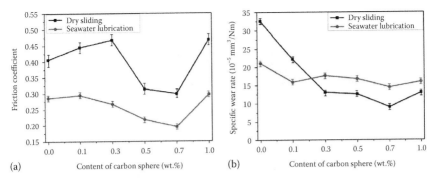

FIGURE 3.24 Variation of friction coefficient (a) and wear rate (b) of PI/carbon sphere microcomposites against the content of carbon sphere under different lubricating conditions Load: 3 N; sliding speed: 0.1569 m/s; duration: 30 minutes. (From Min, C. et al., *Tribol. Inter.*, 90, 175–184, 2015.)

PI/carbon sphere microcomposites have been prepared by the *in situ* polymerization method. The main objective was to investigate the tribological properties of PI/carbon sphere microcomposite films by a universal micro tribotester under dry friction, pure water lubrication, and seawater lubrication conditions. Anti-wear and friction-reducing mechanisms of PI/carbon sphere microcomposites have also been discussed [166]. The friction and wear behaviors of PI/carbon sphere microcomposites under dry friction and seawater lubrication are comparatively investigated as well in Figure 3.24. It is found that irrespective of dry friction or seawater lubrication, 0.7 wt.% PI/carbon sphere microcomposite has the lowest wear rate under the corresponding condition. Figure 3.24a shows that the friction coefficients of PI/carbon sphere microcomposites under seawater lubrication are lower than those under dry friction. Clearly, the 0.7 wt.% PI/carbon sphere microcomposite has the lowest friction coefficient under seawater lubrication. In Figure 3.24b, the wear rate of PI/carbon sphere microcomposite under seawater lubrication changes to be higher than that under dry friction when the content of carbon sphere is more than 0.3 wt.%. The wear rate of 0.7 wt.% PI/carbon sphere microcomposite under seawater lubrication is a little higher than that under dry friction.

The worn surface of PI/carbon sphere microcomposites under dry sliding condition was investigated by optical photographs, as shown in Figure 3.25. There are numerous wide and deep furrows on the worn surface of pure PI, which implies that serious abrasive wear occurs (Figure 3.25a). As carbon sphere is added into the PI matrix, the wide and deep furrows on the worn surface of the PI/carbon sphere microcomposite films are considerably reduced (Figure 3.25b–f). We can see a relatively smooth worn surface without grooves with the increasing content of carbon sphere, which indicates that the wear behavior changes from abrasive wear to adhesive wear. Therefore, under dry friction, PI/carbon sphere microcomposites are easily detached from the surface by wear and then form the transfer film. Owing to the certain amount of transfer film between the surface of the specimen and the counterpart, the damage of worn surface of the PI/carbon sphere microcomposite films decreases with the increasing content of carbon sphere. The incorporation of carbon sphere can greatly improve the tribological performance of PI under dry friction and seawater lubrication.

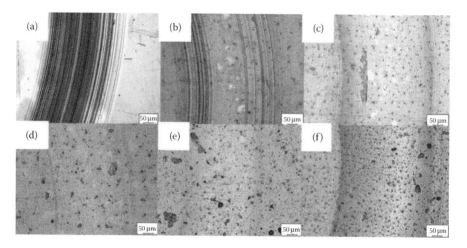

FIGURE 3.25 Optical photographs of the worn surfaces of PI/carbon sphere microcomposites under dry friction: (a) PI, (b) 0.1 wt.% PI/carbon sphere, (c) 0.3 wt.% PI/carbon sphere, (d) 0.5 wt.% PI/carbon sphere, (e) 0.7 wt.% PI/carbon sphere, and (f) 1 wt.% PI/carbon sphere. Load: 3 N; sliding speed: 0.1569 m/s; duration: 30 minutes. (From Min, C. et al., *Tribol. Inter.*, 90, 175–184, 2015.)

This is due to the self-lubricating properties of carbon sphere. However, PI/carbon sphere microcomposite shows the higher wear rate than other conditions because of the fractured surfaces of PI/carbon sphere microcomposites. To begin with, we observe the fractured surface becomes rough and appears some holes with the addition of carbon sphere by SEM. With the increasing loading amount of carbon sphere, compactness of PI/carbon sphere microcomposite decreases. Under dry friction, PI/ carbon sphere microcomposites can be easily detached from the surface by wear and result in the formation of transfer film. The certain amount of transfer film could improve the wear resistance of PI/carbon sphere microcomposite. In addition, during the aqueous lubrication tests, it is likely that water diffuses into the holes of fractured surface regions of composites leading to plasticization, swelling, and softening, resulting in reduction in the hardness and strength, which would cause PI/carbon sphere microcomposite to be less wear resistant under aqueous lubrication. Therefore, with the increasing content of carbon sphere, the wear resistance of PI/carbon sphere microcomposite under dry friction becomes better than that under aqueous lubrication. Further, PI/carbon sphere microcomposite shows higher wear rate under seawater condition than under pure water condition because of corrosion of seawater.

In previous studies, expanded graphite with nanoscale lamellar structure (nano-EG) was prepared. For its self-lubricating properties, the tribological properties of PTFE/nano-EG composite were studied [167]. The research results showed that the addition of nano-EG significantly improved the wear resistance of PTFE composites. In order to expand its application, a research aims to investigate the impact of nano-EG content on the tribological properties of PI-based composites. One of our purposes is to seek a PI-based composite with better self-lubricating properties and high wear resistance, which could be used in rolling or sliding bearings.

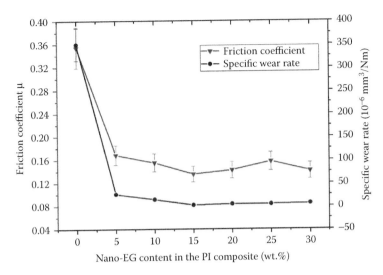

FIGURE 3.26 Frictional coefficient and wear rate of PI/nano-EG composites. (From Jia, Z. et al., *Wear*, 338, 282–287, 2015.)

Figure 3.26 shows the coefficient of friction and wear rate for all the tested PI/nano-EG composites sliding against ring specimens [168]. With the different content of nano-EG, the coefficient of friction of PI/nano-EG composites varies widely. When nano-EG content is zero (i.e., pure PI), the coefficient of friction is 0.354. Even when a small amount of nano-EG is added into PI, the coefficient of friction of PI-based composites is greatly improved. For the PI/nano-EG composite filled with 5 wt.% nano-EG, its coefficient of friction is 0.017, hardly one-twentieth of pure PI. Filled with 10 wt.% nano-EG, its coefficient of friction is 0.155, only 44% of that of pure PI. When nano-EG mass fraction is of 15 wt.%, the coefficient of friction is 0.135 reaching the lowest point. After that, with the increase of nano-EG content in the composites, the coefficient of friction fluctuates in a narrow range. The antifriction property of PI/nano-EG composites strongly depends on the nature of nano-EG because nano-EG itself is a good solid lubricant. Also, from the curve of the wear rate of PI/nano-EG composites, it can be seen that the addition of nano-EG remarkably improves the wear resistance of pure PI. When the filled content of nano-EG is 15 wt.%, the wear rate of PI/nano-EG composite reaches the lowest level (1.8×10^{-6} mm^3/Nm), which increases nearly 200 times compared with that of pure PI (349.7×10^{-6} mm^3/Nm). Compared with the ordinary graphite, the PI composites with nano-EG show better tribological properties [169,170]. As can be seen from the figure, a small amount of nano-EG will be able to make the wear rate of PI drop a lot. With the increase of nano-EG content, the wear rate of PI/nano-EG composites decreases first and then increases in a smaller range. With the gradual increasing of nano-EG, the contact zone between PI/nano-EG composite and 1045 steel matrix covers not only PI matrix but also more nano-EG. The unique self-lubricating of nano-EG makes the antifriction continuously improve. Meanwhile, the good mechanical properties of PI also make the composites still have a relatively high strength and hardness [171], so the PI/nano-EG shows a gradual increase in wear resistance. However, when the nano-EG filler content increases more than 15 wt.%, in PI/nano-EG composites, the wear rate of composites gradually increases with the increasing of nano-EG percentage. This can be explained by the following reasons: first, the increasing content of nano-EG filler means that PI substrate (with good comprehensive mechanical properties) percentage reduces, which leads to an overall performance degradation of PI/nano-EG composite. In addition, the increasing content of filler could exacerbate the nanoparticle agglomeration effect inevitably. Because of the poor mechanics performance of graphite, possible reunion nano-EG evolves into obvious defects (micro-holes), which weaken the mechanical properties of PI/nano-EG composites. Also, the increasing content of filler could decrease adhesion between PI matrix and nano-EG, which would result in worse mechanical properties of composites similarly [171]. For these reasons, under the role of friction force, the wear resistance of PI/nano-EG composites with higher filler content relatively declines. Nevertheless, of all PI/nano-EG composites, the wear rate remains at a low level, which could be resulted from the generation of self-lubricating transfer film with low shear strength.

Figure 3.27 shows the SEM images of the worn surface of PI/nano-EG composite with different nano-EG content. From the micrographs, it can be seen that the addition of nano-EG caused tremendous changes on worn surfaces. When filler content is low, deep furrows of serious adhesion wear and plastic deformation on the worn surface are obviously observed, as seen in Figure 3.27a. With the increasing filler content, the worn surface of PI/nano-EG is smooth and there is a small amount of micro-scratches on the composite surface. These micro-scratches are shallow and fine, whose directions are random. And the wear debris is very tiny, as shown in Figure 3.27b. When nano-EG content exceeds a certain value, the degradation of wear surface morphology is observed (Figure 3.27c), and severe microcrack is found in locally amplified zone, as shown in Figure 3.27d. This phenomenon also occurs in other PI/nano-EG composites with higher nano-EG content, which means that too much nano-EG content weakens the bonding strength between the components of the PI/nano-EG composites, to some extent. However, from the whole, despite higher nano-EG content has some negative effects on PI/nano-EG composites, the self-lubricating property of nano-EG is still able to make PI/nano-EG composites surface be of a very low coefficient of friction. Moreover, because of maintaining sufficient retained compressive strength, PI/nano-EG composites exhibit low wear rate. Furthermore, high content of nano-EG makes its distribution uneven in PI matrix. The agglomerate phenomenon means that around the aggregate nanoparticle stress distortion occurs, which is why obvious microcracks observed on the surface of PI/nano-EG composites happened under high filler content condition. More seriously, pure PI specimen shows obvious characteristics of the adhesive wear and abrasive wear, with deep and intense scratches, as shown in Figure 3.27e. Due to the large shear strength and strong adhesion effect of pure PI, under the action of the surface micro convex bodies on rigid counterface, PI matrix is constantly peeled off to form much yellow PI debris, which is also observed on the specimen surface. Figure 3.28 shows the SEM micrographs of transfer films generated on the surfaces of steel substrate when PI/nano-EG composites filled with different mass fractions of nano-EG slides against 1045 steel substrates under test conditions. As shown in Figure 3.28a, lower filler content of nano-EG brings

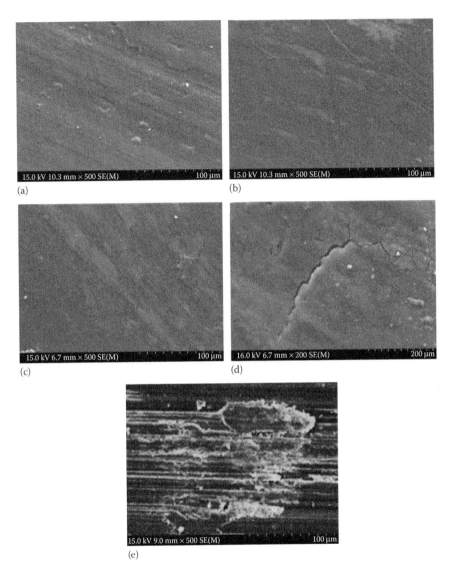

FIGURE 3.27 SEM micrographs of the worn surfaces for PI/nano-EG composites. The nano-EG contents are 10% (a), 15% (b), and 30% (c) mass percentages; (d) pure PI. (From Jia, Z. et al., *Wear*, 338, 282–287, 2015.)

about thin transfer film on steel substrate surface, and the initial machining marks is visible. When the filler mass fraction of nano-EG is 15 wt.%, the transfer film formed on the steel matrix surface is relatively continuous and covers the whole machining surface. The friction marks are shallow and thin, as shown in Figure 3.28b. With the increase of nano-EG content, the transfer film becomes thick, and there are more friction marks and wear debris on the surface, as shown in Figure 3.28c and d. This is well agreed with the previous analysis results. The increasing filler content weakens

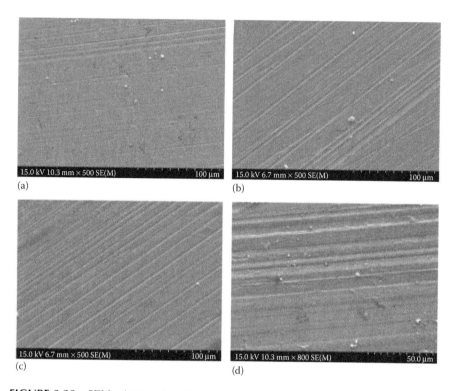

FIGURE 3.28 SEM micrographs of the protective layer formed on the metal matrix. The nano-EG contents are 10% (a), 15% (b), 20% (c), and 30% (d) mass percentages. (From Jia, Z. et al., *Wear*, 338, 282–287, 2015.)

the bonding force between nano-EG and PI resin matrix, and the graphite strength is itself almost zero, so they result in a decline in overall performance of the PI/nano-EG composites. Therefore, if the nano-EG filler content exceeds a threshold, the wear rate begins to increase.

3.7 POLYAMIDE

It is well known that polyamide 6 (PA6) is a good gear and bearing material due to its high strength and good wear resistance [172], but further improvement is still required to meet more demanding applications. Thus, glass fiber [173], nano-Al_2O_3 [174], nano-SiO_2 [175], and clay [176] have been used to improve the mechanical and tribological properties of PA6. CNTs possess unusual mechanical properties and high aspect ratio; therefore, they could be expected to significantly improve the tribological behavior of PA6. However, the research results on the friction and wear behavior of PA6/CNT composites under dry sliding and water-lubricated condition have not been found so far in the literature. A study work would focus on this topic to clarify the effect of CNTs on the friction and wear behavior of PA6 [177]. The variation of the friction coefficient of PA6 and its composites with normal loads under

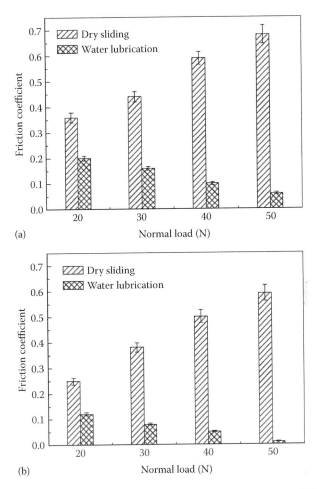

FIGURE 3.29 Effect of the normal load on the friction coefficient of (a) PA6 and (b) its composites under dry sliding and water-lubricated condition.

dry sliding and water-lubricated condition against the stainless steel counterfaces is shown in Figure 3.29. Compared to PA6, PA6 composites exhibited a lower value of friction coefficient under all conditions in this work. This implied that the friction coefficient of PA6 under dry sliding and water-lubricated condition was reduced by the addition of CNTs. From Figure 3.29, it was also found that the effect of normal loads on the friction coefficient of PA6 and its composites under dry sliding and water-lubricated condition exhibited in different ways. Under dry sliding, the friction coefficient of PA6 and its composites increased with the increasing normal load. On the contrary, under water-lubricated condition, the friction coefficient decreased with the increasing normal load. From the previous discussion, it can be seen that CNT was an effective reinforcement for PA6. The PA6/CNT composite had higher tensile strength, Young's modulus, microhardness, crystallinity, and water-absorbing resistance than pure PA6, which therefore resulted in a higher load-carrying capacity of PA6/CNT composites. Under dry sliding, the temperature of the pin surface rose due to the accumulation of frictional heat, which resulted in the rapid increase of the adhesive friction. However, it can be found that the temperature rise of the PA6/CNT composites surface was lower than that of PA6, because CNTs had higher thermal conductivity than PA6 and improved the dissipation of accumulated frictional heat.

The specific wear rates of PA6 and its composites with the increasing normal loads under dry sliding and water-lubricated condition are shown in Figure 3.30. It can be seen that the specific wear rate of PA6 composites always exhibited a lower value than that of PA6 under all conditions in this work, which indicated that the wear resistance of PA6 under dry sliding and water-lubricated condition was improved by the addition of CNTs. Moreover, the specific wear rate of PA6 and its composites under water-lubricated sliding was always higher than that under dry

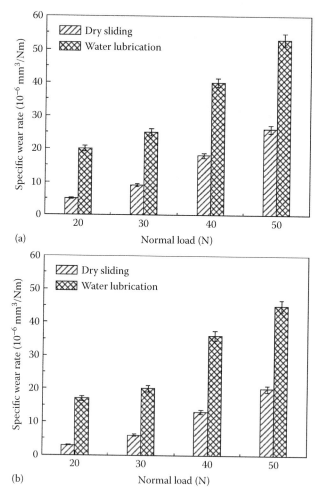

FIGURE 3.30   Effect of the normal load on the specific wear rate of (a) PA6 and (b) its composites under dry sliding and water-lubricated condition.

sliding, which indicated that the distilled water weakened the wear resistance of the two materials. From Figure 3.30, it can also be found that the specific wear rate of both materials exhibited the similar trend with the normal load. Under both dry sliding and water-lubricated condition, the specific wear rate of PA6 and its composites increased with the increasing normal load, which was in consistent with the finding that the wear loss was proportional to the normal load [178]. As results showed, PA6/CNT composites had higher strength, modulus, crystallinity, and microhardness than PA6, which indicated high load-carrying capacity of PA6 composites due to the enhancement of CNTs. Therefore, the removal of the material from PA6 composites was more difficult than that of PA6, which led to a higher wear resistance of PA6 composites under sliding against the stainless steel counterfaces.

Figure 3.31 shows the SEM images of the worn surfaces of PA6 and its composites under dry sliding at a normal load of 50 N. As shown in Figure 3.31a, the worn

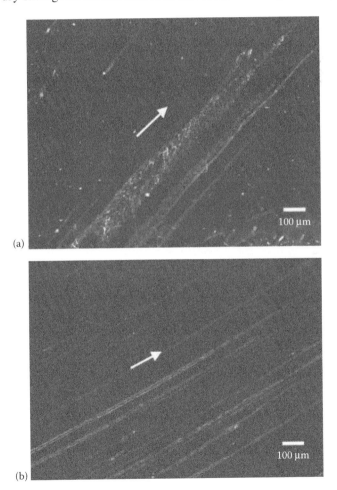

FIGURE 3.31   SEM images of the typical worn surfaces of (a) PA6 and (b) PA6/CNT composites under dry sliding. The sliding direction is shown by the white arrow.

surface of PA6 was very rough, displaying plucked and ploughed marks, which indicated that the wear mechanism was featured with adhesive wear and plough wear. In addition, the worn tracks of PA6 pin were thick and parallel to the sliding direction. By contrast, the ploughing and adhesion on the worn surface of PA6/CNT composites were considerably reduced (Figure 3.31b). A relatively smooth worn surface can be found, indicating that the PA6 reinforced with CNTs was not easily drawn out when sliding against the stainless steel counterfaces. This observation was consistent with the improved wear resistance of PA6/CNT composites. In addition, CNTs that were released from the PA6/CNT composite during sliding might be transferred to the contact zone of PA6 composites and the counterfaces. These CNTs could serve as a solid lubricant to prevent the direct contact between the two mating surfaces, thereby reducing the wear rate and the friction coefficient. Therefore, it can be deduced that the addition of CNTs contributed to restrain the scuffing and adhesion of the PA6 under dry sliding against the stainless steel counterfaces. As a result, the PA6/CNT composites showed much better friction and wear resistance than those of the PA6 matrix.

3.8   POLYSTYRENE

One of the used polymer matrices for the preparation is polystyrene (PS). It is a commonly used polymer with various applications, for example, insulation, packaging, household, and automotive industry. Unfortunately, it also exhibits some disadvantageous properties, that is, relatively high flammability. This is the reason why PS composites and later nanocomposites have been studied because of the possibility of the improvement of PS properties. Excellent properties of PS polymeric material such as its low cost, mechanical robustness [179], readily availability, and fine processability make it an ideal candidate to be used for composite [180,181]. PS is an important commodity plastic; however, it is not known as a suitable material for tribological applications. It has a relatively high friction coefficient in contact with metallic surfaces and high abrasion loss factor.

The basic structure of MoS_2 is organized in two layers of sulfur atoms forming a sandwich structure, with a layer of molybdenum atoms in the middle. Due to the weak van der Waals interactions between the sheets, MoS_2 shows a low friction coefficient and thus gives rise to its superior lubricating properties [182,183]. A study was conducted to prepare and characterize MoS_2-based composites with a commodity polymer such as PS. Thermal stability, wettability, mechanical, and tribological properties of the resulting materials were investigated [184]. As presented in Table 3.1, COF exhibits the highest reduction (61%) with respect to PS when it is reinforced with 0.75 wt.% of MoS_2 (PS-MoS_2-OA-3). The COF has a decrease of 57% for PS-MoS_2-OA-2 and only 7% for PS-MoS_2-OA-1. Analyzing the Table 3.1, it is evident that there is an increment in the wear track depth for each specimen compared to the PS sample. The highest value is found for the PS-MoS_2-OA-3, which exhibited a surprising large wear track dimension. The PS-MoS_2-OA-1 and PS-MoS_2-OA-2 are also more worn than the PS sample, presenting deeper wear track. The increasing in sample rigidity (elastic modulus) seems to decrease the erosion volume. Similar

TABLE 3.1

Elastic Modulus of PS and Composite Samples, Average Friction Coefficient MoS₂ Oleylamine Content, Wear Track Depth for the Different Samples

Sample Name	Inorganic Filler Content	50% Degree Temperature (°C)	Storage Modulus (MPa)			Friction Coefficient	Wear Track Depth (µm)
			$T = 25°C$	$T = 40°C$	$T = 80°C$		
PS	0	407	1909	1930	1071	0.72	7
PS-MoS₂-OA-1	0.3%	415	1978	1978	838	0.67	17
PS-MoS₂-OA-2	1.6%	420	2072	1918	30	0.31	56
PS-MoS₂-OA-3	4.5%	424	1734	1574	13	0.28	83

results were found by Yu et al. [185] and Wang et al. [186]. They have shown that MoS_2 deteriorated the wear resistance of PPS, whereas graphite and PTFE contributed to its increase. The negative effect was attributed to poor mechanical properties and a tendency of the MoS_2 to segregate and extrude out of the matrix material in the process of sliding. In both studies, the additive concentration in the lubricant was very high (higher than 10 vol.%). At these concentrations of MoS_2, a significant reduction of the mechanical strength was observed for the majority of polymer composites.

REFERENCES

1. Bhushan B. *Modern Tribology Handbook*, Two Volume Set. CRC Press: Boca Raton, FL; 2000.
2. Seabra LsC, Baptista AM. Tribological behaviour of food grade polymers against stainless steel in dry sliding and with sugar. *Wear.* 2002;253:394–402.
3. Scharf T, Prasad S. Solid lubricants: A review. *Journal of Materials Science.* 2013;48:511–531.
4. Burris DL, Sawyer WG. A low friction and ultra low wear rate PEEK/PTFE composite. *Wear.* 2006;261:410–418.
5. Burris DL, Sawyer WG. Improved wear resistance in alumina-PTFE nanocomposites with irregular shaped nanoparticles. *Wear.* 2006;260:915–918.
6. McCook NL, Boesl B, Burris DL, Sawyer WG. Epoxy, ZnO, and PTFE nanocomposite: Friction and wear optimization. *Tribology Letters.* 2006;22:253–257.
7. Lancaster JK. Polymer-based bearing materials: The role of fillers and fibre reinforcement. *Tribology.* 1972;5:249–255.
8. Blanchet TA, Kennedy FE. Sliding wear mechanism of polytetrafluoroethylene (PTFE) and PTFE composites. *Wear.* 1992;153:229–243.
9. Ye J, Moore AC, Burris DL. Transfer film tenacity: A case study using ultra-low-wear alumina–PTFE. *Tribology Letters.* 2015;59:1–11.
10. Burris DL. Wear-resistance mechanisms in polytetrafluoroethylene (PTFE) based tribological nanocomposites, University of Florida: Gainesville, FL; 2006.
11. Burris DL, Zhao S, Duncan R, Lowitz J, Perry SS, Schadler LS et al. A route to wear resistant PTFE via trace loadings of functionalized nanofillers. *Wear.* 2009;267:653–660.

12. Krick BA, Ewin JJ, Blackman GS, Junk CP, Gregory Sawyer W. Environmental dependence of ultra-low wear behavior of polytetrafluoroethylene (PTFE) and alumina composites suggests tribochemical mechanisms. *Tribology International*. 2012;51:42–46.

13. Pitenis AA, Ewin JJ, Harris KL, Sawyer WG, Krick BA. In vacuo tribological behavior of polytetrafluoroethylene (PTFE) and alumina nanocomposites: The importance of water for ultralow wear. *Tribology Letters*. 2014;53(1):189–197.

14. Lancaster JK. Lubrication by transferred films of solid lubricants. *ASLE Transactions*. 1965;8:146–155.

15. Briscoe BJ, Pogosian AK, Tabor D. The friction and wear of high density polythene: The action of lead oxide and copper oxide fillers. *Wear*. 1974;27:19–34.

16. Briscoe B. Wear of polymers: An essay on fundamental aspects. *Tribology International*. 1981;14:231–243.

17. Bahadur S, Gong D, Anderegg JW. The role of copper compounds as fillers in transfer film formation and wear of nylon. *Wear*. 1992;154:207–223.

18. Briscoe BJ, Sinha SK. Wear of polymers. *Proceedings of the Institution of Mechanical Engineers, Part J: Journal of Engineering Tribology*. 2002;216:401–413.

19. Bahadur S, Sunkara C. Effect of transfer film structure, composition and bonding on the tribological behavior of polyphenylene sulfide filled with nano particles of TiO_2, ZnO, CuO and SiC. *Wear*. 2005;258:1411–1421.

20. Friedrich K, Zhang Z, Schlarb AK. Effects of various fillers on the sliding wear of polymer composites. *Composites Science and Technology*. 2005;65:2329–2343.

21. McCook NL, Burris DL, Bourne GR, Steffens J, Hanrahan JR, Sawyer WG. Wear resistant solid lubricant coating made from PTFE and epoxy. *Tribology Letters*. 2005;18:119–124.

22. Burris DL, Boesl B, Bourne GR, Sawyer WG. Polymeric nanocomposites for tribological applications. *Macromolecular Materials and Engineering*. 2007;292:387–402.

23. Ye J, Burris DL, Xie T. A review of transfer films and their role in ultra-low-wear sliding of polymers. *Lubricants*. 2016;4:4.

24. Ojha S, Acharya SK, Raghavendra G. Mechanical properties of natural carbon black reinforced polymer composites. *Journal of Applied Polymer Science*. 2015;132. DOI: 10.1002/app.41211.

25. Rabby M, Jeelani S, Rangari VK. Microwave processing of SiC nanoparticles infused polymer composites: Comparison of thermal and mechanical properties. *Journal of Applied Polymer Science*. 2015;132. DOI: 10.1002/app.4170.

26. Barari B, Omrani E, Moghadam AD, Menezes PL, Pillai KM, Rohatgi PK. Mechanical, physical and tribological characterization of nano-cellulose fibers reinforced bio-epoxy composites: an attempt to fabricate and scale the 'Green' composite. *Carbohydrate Polymers*. 2016;147:282–293.

27. Omrani E, Barari B, Moghadam AD, Rohatgi PK, Pillai KM. Mechanical and tribological properties of self-lubricating bio-based carbon-fabric epoxy composites made using liquid composite molding. *Tribology International*. 2015;92:222–232.

28. Zhao Q, Bahadur S. Investigation of the transition state in the wear of polyphenylene sulfide sliding against steel. *Tribology Letters*. 2002;12:23–33.

29. Schönherr H, Vancso GJ. The mechanism of PTFE and PE friction deposition: A combined scanning electron and scanning force microscopy study on highly oriented polymeric sliders. *Polymer*. 1998;39:5705–5709.

30. Partridge IK. *Advanced Composites*. Elsevier: London, UK; 1989.

31. Harris B. *Engineering Composite Materials*. The Institute of Metals: London, UK; 1986.

32. Friedrich K, Lu Z, Hager A. Recent advances in polymer composites' tribology. *Wear*. 1995;190:139–144.

33. Friedrich K, Zhang Z, Schlarb AK. Effects of various fillers on the sliding wear of polymer composites. *Composites Science and Technology*. 2005;65:2329–2343.

34. Fusaro RL. Self-lubricating polymer composites and polymer transfer film lubrication for space applications. *Tribology International.* 1990;23:105–122.
35. Sandler J, Shaffer M, Prasse T, Bauhofer W, Schulte K, Windle A. Development of a dispersion process for carbon nanotubes in an epoxy matrix and the resulting electrical properties. *Polymer.* 1999;40:5967–5971.
36. Schadler L, Giannaris S, Ajayan P. Load transfer in carbon nanotube epoxy composites. *Applied Physics Letters.* 1998;73:3842–3844.
37. Cooper C, Young R, Halsall M. Mechanical properties of carbon nanotube. *Composites Part A: Applied Science and Manufacturing.* 2000;32:401.
38. Puglia D, Valentini L, Kenny J. Analysis of the cure reaction of carbon nanotubes/epoxy resin composites through thermal analysis and Raman spectroscopy. *Journal of Applied Polymer Science.* 2003;88:452–458.
39. Bennett S, Johnson D. Structural heterogeneity in carbon fibers. *Proceedings of the Fifth London International Carbon and Graphite Conference.* Society of Chemical Industry: London; 1978. p. 377.
40. Soule D, Nezbeda C. Direct basal-plane shear in single-crystal graphite. *Journal of Applied Physics.* 1968;39:5122–5139.
41. Nak-Ho S, Suh NP. Effect of fiber orientation on friction and wear of fiber reinforced polymeric composites. *Wear.* 1979;53:129–141.
42. Stachowiak G, Batchelor AW. *Engineering Tribology.* Butterworth-Heinemann: Oxford, UK; 2013.
43. Suh NP, Sin H-C. The genesis of friction. *Wear.* 1981;69:91–114.
44. Larsen TØ, Andersen TL, Thorning B, Vigild ME. The effect of particle addition and fibrous reinforcement on epoxy-matrix composites for severe sliding conditions. *Wear.* 2008;264:857–868.
45. Basavarajappa S, Ellangovan S, Arun K. Studies on dry sliding wear behaviour of graphite filled glass–epoxy composites. *Materials & Design.* 2009;30:2670–2675.
46. Zhi RM, Qiu ZM, Liu H, Zeng H, Wetzel B, Friedrich K. Microstructure and tribological behavior of polymeric nanocomposites. *Industrial Lubrication and Tribology.* 2001;53:72–77.
47. Ng C, Schadler L, Siegel R. Synthesis and mechanical properties of TiO_2–epoxy nanocomposites. *Nanostructured Materials.* 1999;12:507–510.
48. Yu L, Yang S, Wang H, Xue Q. An investigation of the friction and wear behaviors of micrometer copper particle- and nanometer copper particle-filled polyoxymethylene composites. *Journal of Applied Polymer Science.* 2000;77:2404–2410.
49. Xue Q-J, Wang Q-H. Wear mechanisms of polyetheretherketone composites filled with various kinds of SiC. *Wear.* 1997;213:54–58.
50. Zhang MQ, Rong MZ, Yu SL, Wetzel B, Friedrich K. Effect of particle surface treatment on the tribological performance of epoxy based nanocomposites. *Wear.* 2002;253:1086–1093.
51. Guo QB, Rong MZ, Jia GL, Lau KT, Zhang MQ. Sliding wear performance of nano-SiO_2/short carbon fiber/epoxy hybrid composites. *Wear.* 2009;266:658–665.
52. Briscoe B. The tribology of composite materials: A preface. In: Friedrich K, (Ed.). *Advances in Composite Technology.* Elsevier: Amsterdam, the Netherlands; 1993. 3–15.
53. Bonfield W, Edwards B, Markham A, White J. Wear transfer films formed by carbon fibre reinforced epoxy resin sliding on stainless steel. *Wear.* 1976;37:113–121.
54. Zhang MQ, Rong MZ, Yu SL, Wetzel B, Friedrich K. Improvement of tribological performance of epoxy by the addition of irradiation grafted nano-inorganic particles. *Macromolecular Materials and Engineering.* 2002;287:111–115.
55. Kadiyala AK, Bijwe J. Surface lubrication of graphite fabric reinforced epoxy composites with nano- and micro-sized hexagonal boron nitride. *Wear.* 2013;301:802–809.

56. Dong B, Yang Z, Huang Y, Li H-L. Study on tribological properties of multi-walled carbon nanotubes/epoxy resin nanocomposites. *Tribology Letters.* 2005;20:251–254.

57. Chen W, Tu J, Wang L, Gan H, Xu Z, Zhang X. Tribological application of carbon nanotubes in a metal-based composite coating and composites. *Carbon.* 2003;41:215–222.

58. Cai H, Yan F, Xue Q. Investigation of tribological properties of polyimide/carbon nanotube nanocomposites. *Materials Science and Engineering: A.* 2004;364:94–100.

59. Chen W, Li F, Han G, Xia J, Wang L, Tu J et al. Tribological behavior of carbon-nanotube-filled PTFE composites. *Tribology Letters.* 2003;15:275–278.

60. Chen W, Tu J, Xu Z, Chen W, Zhang X, Cheng D. Tribological properties of Ni–P-multi-walled carbon nanotubes electroless composite coating. *Materials Letters.* 2003;57:1256–1260.

61. Lu M, Wang Z, Li H-L, Guo X-Y, Lau K-T. Formation of carbon nanotubes in silicon-coated alumina nanoreactor. *Carbon.* 2004;42:1846–1849.

62. Zhao G, Hussainova I, Antonov M, Wang Q, Wang T. Friction and wear of fiber reinforced polyimide composites. *Wear.* 2013;301:122–129.

63. Feng X, Wang H, Shi Y, Chen D, Lu X. The effects of the size and content of potassium titanate whiskers on the properties of PTW/PTFE composites. *Materials Science and Engineering: A.* 2007;448:253–258.

64. Feng X, Diao X, Shi Y, Wang H, Sun S, Lu X. A study on the friction and wear behavior of polytetrafluoroethylene filled with potassium titanate whiskers. *Wear.* 2006;261:1208–1212.

65. Kandanur SS, Rafiee MA, Yavari F, Schrameyer M, Yu Z-Z, Blanchet TA et al. Suppression of wear in graphene polymer composites. *Carbon.* 2012;50:3178–3183.

66. Zhu J, Shi Y, Feng X, Wang H, Lu X. Prediction on tribological properties of carbon fiber and TiO_2 synergistic reinforced polytetrafluoroethylene composites with artificial neural networks. *Materials & Design.* 2009;30:1042–1049.

67. Mu L, Chen J, Shi Y, Feng X, Zhu J, Wang H et al. Durable polytetrafluoroethylene composites in harsh environments: Tribology and corrosion investigation. *Journal of Applied Polymer Science.* 2012;124:4307–4314.

68. Shi Y, Feng X, Wang H, Lu X. The effect of surface modification on the friction and wear behavior of carbon nanofiber-filled PTFE composites. *Wear.* 2008;264:934–939.

69. Shi Y, Mu L, Feng X, Lu X. The tribological behavior of nanometer and micrometer TiO_2 particle-filled polytetrafluoroethylene/polyimide. *Materials & Design.* 2011;32:964–970.

70. Villavicencio M, Renouf M, Saulot A, Michel Y, Mahéo Y, Colas G et al. Self-lubricating composite bearings: Effect of fibre length on its tribological properties by dem modelling. *Tribology International.* 2016.

71. Klaas N, Marcus K, Kellock C. The tribological behaviour of glass filled polytetrafluoroethylene. *Tribology International.* 2005;38:824–833.

72. Voss H, Friedrich K. On the wear behaviour of short-fibre-reinforced PEEK composites. *Wear.* 1987;116:1–18.

73. Li F, Hu K-A, Li J-L, Zhao B-Y. The friction and wear characteristics of nanometer ZnO filled polytetrafluoroethylene. *Wear.* 2001;249:877–882.

74. Mcelwain SE, Blanchet TA, Schadler LS, Sawyer WG. Effect of particle size on the wear resistance of alumina-filled PTFE micro-and nanocomposites. *Tribology Transactions.* 2008;51:247–253.

75. Lai SQ, Li TS, Liu XJ, Lv RG. A study on the friction and wear behavior of PTFE filled with acid treated nano-attapulgite. *Macromolecular Materials and Engineering.* 2004;289:916–922.

76. Blanchet TA, Kandanur SS, Schadler LS. Coupled effect of filler content and countersurface roughness on PTFE nanocomposite wear resistance. *Tribology Letters.* 2010;40:11–21.
77. Lu Z, Friedrich K. On sliding friction and wear of PEEK and its composites. *Wear.* 1995;181:624–631.
78. Zhang Z, Breidt C, Chang L, Friedrich K. Wear of PEEK composites related to their mechanical performances. *Tribology International.* 2004;37:271–277.
79. Cirino M, Friedrich K, Pipes R. The effect of fiber orientation on the abrasive wear behavior of polymer composite materials. *Wear.* 1988;121:127–141.
80. Friedrich K. Wear model for multiphase materials and synergistic effect in polymeric hybrid composites. In: Friedrich K, Pipes RB, (Eds.). *Advances in Composite Technology, Composite Materials Series.* Elsevier: Amsterdam, the Netherlands; 1993. 209–273.
81. Zhang G, Schlarb A. Correlation of the tribological behaviors with the mechanical properties of poly-ether-ether-ketones (PEEKs) with different molecular weights and their fiber filled composites. *Wear.* 2009;266:337–344.
82. Chang L, Zhang Z, Breidt C, Friedrich K. Tribological properties of epoxy nanocomposites: I. Enhancement of the wear resistance by nano-TiO$_2$ particles. *Wear.* 2005;258:141–148.
83. Zhang G, Rasheva Z, Schlarb A. Friction and wear variations of short carbon fiber (SCF)/PTFE/graphite (10vol.%) filled PEEK: Effects of fiber orientation and nominal contact pressure. *Wear.* 2010;268:893–899.
84. Chand S. Review carbon fibers for composites. *Journal of Materials Science.* 2000;35:1303–1313.
85. Cao L, Shen XJ, Li RY. Three-dimensional thermal analysis of spherical plain bearings with self-lubricating fabric liner. *Advanced Materials Research.* 2010;97–101:3366–3370.
86. Wielage B, Müller T, Lampke T. Design of ceramic high-accuracy bearings containing textile fabrics. *Materialwissenschaft Und Werkstofftechnik.* 2007;38:79–84.
87. Qiu M, Gao Z, Yao S, Chen L. Effects of oscillation frequency on the tribological properties of self-lubrication spherical plain bearings with PTFE woven liner. *Key Engineering Materials.* 2011;455:406–410.
88. Rattan R, Bijwe J. Carbon fabric reinforced polyetherimide composites: Influence of weave of fabric and processing parameters on performance properties and erosive wear. *Materials Science and Engineering: A.* 2006;420:342–350.
89. Park DC, Lee SM, Kim BC, Kim HS. Development of heavy duty hybrid carbon–phenolic hemispherical bearings. *Composite Structures.* 2006;73:88–98.
90. Kim SS, Yu HN, Hwang IU, Kim SN, Suzuki K, Sada H. The sliding friction of hybrid composite journal bearing under various test conditions. *Tribology Letters.* 2009;35:211–219.
91. Lancaster J. Accelerated wear testing of PTFE composite bearing materials. *Tribology International.* 1979;12:65–75.
92. Azizi SMAS, Alloin F, Dufresne A. Review of recent research into cellulosic whiskers, their properties and their application in nanocomposite field. *Biomacromolecules.* 2005;6:612–626.
93. Moon RJ, Martini A, Nairn J, Simonsen J, Youngblood J. Cellulose nanomaterials review: Structure, properties and nanocomposites. *Chemical Society Reviews.* 2011;40:3941–3994.
94. Hussain F, Hojjati M, Okamoto M, Gorga RE. Review article: Polymer-matrix nanocomposites, processing, manufacturing, and application: An overview. *Journal of Composite Materials.* 2006;40:1511–1575.

95. Kurahatti R, Surendranathan A, Kori S, Singh N, Kumar AR, Srivastava S. Defence applications of polymer nanocomposites. *Defence Science Journal.* 2010;60(5):551–563.

96. Briscoe BJ, Sinha SK. Tribological applications of polymers and their composites: past, present and future prospects. *Tribology and Interface Engineering Series.* 2008;55:1–14.

97. Winey KI, Vaia RA. Polymer nanocomposites. *MRS Bulletin.* 2007;32:314–322.

98. Ray SS, Okamoto M. Polymer/layered silicate nanocomposites: A review from preparation to processing. *Progress in Polymer Science.* 2003;28:1539–1641.

99. Pavlidou S, Papaspyrides C. A review on polymer–layered silicate nanocomposites. *Progress in Polymer Science.* 2008;33:1119–1198.

100. Yu Y, Gu J, Kang F, Kong X, Mo W. Surface restoration induced by lubricant additive of natural minerals. *Applied Surface Science.* 2007;253:7549–7553.

101. Pogodaev L, Buyanovskii I, Kryukov EY, Kuz'min V, Usachev V. The mechanism of interaction between natural laminar hydrosilicates and friction surfaces. *Journal of Machinery Manufacture and Reliability.* 2009;38:476.

102. Fan B, Yang Y, Feng C, Ma J, Tang Y, Dong Y et al. Tribological properties of fabric self-lubricating liner based on organic montmorillonite (OMMT) reinforced phenolic (PF) nanocomposites as hybrid matrices. *Tribology Letters.* 2015;57:22.

103. Chang IT, Sancaktar E. Clay dispersion effects on excimer laser ablation of polymer–clay nanocomposites. *Journal of Applied Polymer Science.* 2013;130:2336–2344.

104. Wang CC, Chen CC. Some physical properties of various amine-pretreated Nomex Aramid yarns. *Journal of Applied Polymer Science.* 2005;96:70–76.

105. Reis P, Ferreira J, Santos P, Richardson M, Santos J. Impact response of Kevlar composites with filled epoxy matrix. *Composite Structures.* 2012;94:3520–3528.

106. Gu H. Research on thermal properties of Nomex/Viscose FR fibre blended fabric. *Materials & Design.* 2009;30:4324–4327.

107. Kim SH, Seong JH, Oh KW. Effect of dopant mixture on the conductivity and thermal stability of polyaniline/nomex conductive fabric. *Journal of Applied Polymer Science.* 2002;83:2245–2254.

108. Schultz GR. Energy weapon protection fabric. U.S. Patent No 8,132,597. 2011.

109. Schultz GR. Energy weapon protection fabric. U.S. Patent No 8,001,999. 2012.

110. Lopes C, Tschopp R. Titanium spherical plain bearing with liner and treated surface. U.S. Patent No EP 1837534 A3. 2007.

111. Ren G, Zhang Z, Zhu X, Ge B, Guo F, Men X et al. Influence of functional graphene as filler on the tribological behaviors of Nomex fabric/phenolic composite. *Composites Part A: Applied Science and Manufacturing.* 2013;49:157–164.

112. Su F-H, Zhang Z-Z, Wang K, Liu W-M. Friction and wear of Synfluo 180XF wax and nano-Al_2O_3 filled Nomex fabric composites. *Materials Science and Engineering: A.* 2006;430:307–313.

113. Su F-H, Zhang Z-Z, Liu W-M. Tribological and mechanical properties of Nomex fabric composites filled with polyfluo 150 wax and nano-SiO_2. *Composites Science and Technology.* 2007;67:102–110.

114. Leven I, Krepel D, Shemesh O, Hod O. Robust superlubricity in graphene/h-BN heterojunctions. *The Journal of Physical Chemistry Letters.* 2012;4:115–120.

115. Lee H, Lee N, Seo Y, Eom J, Lee S. Comparison of frictional forces on graphene and graphite. *Nanotechnology.* 2009;20:325701.

116. Kim K-S, Lee H-J, Lee C, Lee S-K, Jang H, Ahn J-H et al. Chemical vapor deposition-grown graphene: The thinnest solid lubricant. *ACS Nano.* 2011;5:5107–5114.

117. Lin L-Y, Kim D-E, Kim W-K, Jun S-C. Friction and wear characteristics of multilayer graphene films investigated by atomic force microscopy. *Surface and Coatings Technology.* 2011;205:4864–4869.

118. Hu J, Jo S, Ren Z, Voevodin A, Zabinski J. Tribological behavior and graphitization of carbon nanotubes grown on 440C stainless steel. *Tribology Letters*. 2005;19:119–125.

119. Pan B, Xu G, Zhang B, Ma X, Li H, Zhang Y. Preparation and tribological properties of polyamide 11/graphene coatings. *Polymer-Plastics Technology and Engineering*. 2012;51:1163–1166.

120. Pan B, Zhao J, Zhang Y, Zhang Y. Wear performance and mechanisms of polyphenylene sulfide/polytetrafluoroethylene wax composite coatings reinforced by graphene. *Journal of Macromolecular Science, Part B*. 2012;51:1218–1227.

121. Du J, Zhao L, Zeng Y, Zhang L, Li F, Liu P et al. Comparison of electrical properties between multi-walled carbon nanotube and graphene nanosheet/high density polyethylene composites with a segregated network structure. *Carbon*. 2011;49:1094–1100.

122. Liu W-W, Yan X-B, Lang J-W, Peng C, Xue Q-J. Flexible and conductive nanocomposite electrode based on graphene sheets and cotton cloth for supercapacitor. *Journal of Materials Chemistry*. 2012;22:17245–17253.

123. Fang M, Wang K, Lu H, Yang Y, Nutt S. Covalent polymer functionalization of graphene nanosheets and mechanical properties of composites. *Journal of Materials Chemistry*. 2009;19:7098–7105.

124. Zhang HJ, Zhang ZZ, Guo F. Tribological behaviors of hybrid PTFE/Nomex fabric/phenolic composite reinforced with multiwalled carbon nanotubes. *Journal of Applied Polymer Science*. 2012;124:235–241.

125. Ren G, Zhang Z, Zhu X, Men X, Jiang W, Liu W. Tribological behaviors of hybrid PTFE/nomex fabric/phenolic composite under dry and water-bathed sliding conditions. *Tribology Transactions*. 2014;57:1116–1121.

126. Yen HJ, Chen CJ, Liou GS. Flexible multi-colored electrochromic and volatile polymer memory devices derived from starburst triarylamine-based electroactive polyimide. *Advanced Functional Materials*. 2013;23:5307–5316.

127. Lin L, Wang A, Dong M, Zhang Y, He B, Li H. Sulfur removal from fuel using zeolites/polyimide mixed matrix membrane adsorbents. *Journal of Hazardous Materials*. 2012;203:204–212.

128. Lu N, Lu C, Yang S, Rogers J. Highly sensitive skin-mountable strain gauges based entirely on elastomers. *Advanced Functional Materials*. 2012;22:4044–4050.

129. Cao L, Sun Q, Wang H, Zhang X, Shi H. Enhanced stress transfer and thermal properties of polyimide composites with covalent functionalized reduced graphene oxide. *Composites Part A: Applied Science and Manufacturing*. 2015;68:140–148.

130. Kwon J, Kim J, Lee J, Han P, Park D, Han H. Fabrication of polyimide composite films based on carbon black for high-temperature resistance. *Polymer Composites*. 2014;35:2214–2220.

131. Jiang Q, Wang X, Zhu Y, Hui D, Qiu Y. Mechanical, electrical and thermal properties of aligned carbon nanotube/polyimide composites. *Composites Part B: Engineering*. 2014;56:408–412.

132. Mu L, Shi Y, Feng X, Zhu J, Lu X. The effect of thermal conductivity and friction coefficient on the contact temperature of polyimide composites: Experimental and finite element simulation. *Tribology International*. 2012;53:45–52.

133. Samyn P, Schoukens G. Thermochemical sliding interactions of short carbon fiber polyimide composites at high pv-conditions. *Materials Chemistry and Physics*. 2009;115:185–195.

134. Huang T, Xin Y, Li T, Nutt S, Su C, Chen H et al. Modified graphene/polyimide nanocomposites: Reinforcing and tribological effects. *ACS Applied Materials & Interfaces*. 2013;5:4878–4891.

135. Gofman I, Zhang B, Zang W, Zhang Y, Song G, Chen C et al. Specific features of creep and tribological behavior of polyimide-carbon nanotubes nanocomposite films: Effect of the nanotubes functionalization. *Journal of Polymer Research*. 2013;20:258.

136. Samyn P, De Baets P, Schoukens G. Influence of internal lubricants (PTFE and silicon oil) in short carbon fibre-reinforced polyimide composites on performance properties. *Tribology Letters.* 2009;36:135–146.

137. Samyn P, Schoukens G. Tribological properties of PTFE-filled thermoplastic polyimide at high load, velocity, and temperature. *Polymer Composites.* 2009;30:1631–1646.

138. Jia J, Zhou H, Gao S, Chen J. A comparative investigation of the friction and wear behavior of polyimide composites under dry sliding and water-lubricated condition. *Materials Science and Engineering: A.* 2003;356:48–53.

139. Jia J, Chen J, Zhou H, Hu L, Chen L. Comparative investigation on the wear and transfer behaviors of carbon fiber reinforced polymer composites under dry sliding and water lubrication. *Composites Science and Technology.* 2005;65:1139–1147.

140. Zhang X-R, Pei X-Q, Wang Q-H. Friction and wear studies of polyimide composites filled with short carbon fibers and graphite and micro SiO_2. *Materials & Design.* 2009;30:4414–4420.

141. Yijun S, Liwen M, Xin F, Xiaohua L. Tribological behavior of carbon nanotube and polytetrafluoroethylene filled polyimide composites under different lubricated conditions. *Journal of Applied Polymer Science.* 2011;121:1574–1578.

142. Liu B, Pei X, Wang Q, Sun X, Wang T. Effects of atomic oxygen irradiation on structural and tribological properties of polyimide/Al_2O_3 composites. *Surface and Interface Analysis.* 2012;44:372–376.

143. Liu H, Wang T, Wang Q. Tribological properties of thermosetting polyimide/TiO_2 nanocomposites under dry sliding and water-lubricated conditions. *Journal of Macromolecular Science, Part B.* 2012;51:2284–2296.

144. Rajesh CC, Ravikumar T. Mechanical and three-body abrasive wear behaviour of Nano-Flyash/ZrO2 filled polyimide composites. *International Journal of Science Research.* 2013;1:196–202.

145. Hanemann T, Szabó DV. Polymer-nanoparticle composites: From synthesis to modern applications. *Materials.* 2010;3:3468–3517.

146. Ye J, Khare H, Burris D. Transfer film evolution and its role in promoting ultra-low wear of a PTFE nanocomposite. *Wear.* 2013;297:1095–1102.

147. Chang L, Friedrich K, Ye L. Study on the transfer film layer in sliding contact between polymer composites and steel disks using nanoindentation. *Journal of Tribology.* 2014;136:021602.

148. Wang Q, Zhang X, Pei X. Study on the synergistic effect of carbon fiber and graphite and nanoparticle on the friction and wear behavior of polyimide composites. *Materials & Design.* 2010;31:3761–3768.

149. Zhang G, Schlarb A, Tria S, Elkedim O. Tensile and tribological behaviors of PEEK/nano-SiO_2 composites compounded using a ball milling technique. *Composites Science and Technology.* 2008;68:3073–3080.

150. Li X, Gao Y, Xing J, Wang Y, Fang L. Wear reduction mechanism of graphite and MoS_2 in epoxy composites. *Wear.* 2004;257:279–283.

151. Vail J, Krick B, Marchman K, Sawyer WG. Polytetrafluoroethylene (PTFE) fiber reinforced polyetheretherketone (PEEK) composites. *Wear.* 2011;270:737–741.

152. Omar MF, Akil HM, Ahmad ZA, Mahmud S. The effect of loading rates and particle geometry on compressive properties of polypropylene/zinc oxide nanocomposites: Experimental and numerical prediction. *Polymer Composites.* 2012;33:99–108.

153. Chang BP, Akil HM, Nasir RBM, Bandara I, Rajapakse S. The effect of ZnO nanoparticles on the mechanical, tribological and antibacterial properties of ultra-high molecular weight polyethylene. *Journal of Reinforced Plastics and Composites.* 2014;33:674–686.

154. Mu L, Zhu J, Fan J, Zhou Z, Shi Y, Feng X et al. Self-lubricating polytetrafluoroethylene/polyimide blends reinforced with zinc oxide nanoparticles. *Journal of Nanomaterials.* 2015;16:373.

155. Díez-Pascual AM, Xu C, Luque R. Development and characterization of novel poly (ether ether ketone)/ZnO bionanocomposites. *Journal of Materials Chemistry B*. 2014;2:3065–3078.
156. Shi Y, Feng X, Wang H, Lu X, Shen J. Tribological and mechanical properties of carbon-nanofiber-filled polytetrafluoroethylene composites. *Journal of Applied Polymer Science*. 2007;104:2430–2437.
157. Wang J, Feng S, Song Y, Li W, Gao W, Elzatahry AA et al. Synthesis of hierarchically porous carbon spheres with yolk-shell structure for high performance supercapacitors. *Catalysis Today*. 2015;243:199–208.
158. Wang Y, Su F, Wood CD, Lee JY, Zhao XS. Preparation and characterization of carbon nanospheres as anode materials in lithium-ion secondary batteries. *Industrial & Engineering Chemistry Research*. 2008;47:2294–2300.
159. Tang S, Tang Y, Vongehr S, Zhao X, Meng X. Nanoporous carbon spheres and their application in dispersing silver nanoparticles. *Applied Surface Science*. 2009;255:6011–6016.
160. Demir-Cakan R, Makowski P, Antonietti M, Goettmann F, Titirici M-M. Hydrothermal synthesis of imidazole functionalized carbon spheres and their application in catalysis. *Catalysis Today*. 2010;150:115–118.
161. Tang S, Vongehr S, Meng X. Carbon spheres with controllable silver nanoparticle doping. *The Journal of Physical Chemistry C*. 2009;114:977–982.
162. Zhang J, Zhang Y, Lian S, Liu Y, Kang Z, Lee S-T. Highly ordered macroporous carbon spheres and their catalytic application for methanol oxidation. *Journal of Colloid and Interface Science*. 2011;361:503–508.
163. Xiong H, Motchelaho MA, Moyo M, Jewell LL, Coville NJ. Correlating the preparation and performance of cobalt catalysts supported on carbon nanotubes and carbon spheres in the Fischer–Tropsch synthesis. *Journal of Catalysis*. 2011;278:26–40.
164. Bhushan B, Gupta B, Van CGW, Capp C, Coe JV. Fullerene (C60) films for solid lubrication. *Tribology Transactions*. 1993;36:573–580.
165. Pozdnyakov A, Kudryavtsev V, Friedrich K. Sliding wear of polyimide-C_{60} composite coatings. *Wear*. 2003;254:501–513.
166. Min C, Nie P, Tu W, Shen C, Chen X, Song H. Preparation and tribological properties of polyimide/carbon sphere microcomposite films under seawater condition. *Tribology International*. 2015;90:175–184.
167. Yang Y-L, Jia Z-N, Chen J-J, Fan B-L. Tribological behaviors of PTFE-based composites filled with nanoscale lamellar structure expanded graphite. *Journal of Tribology*. 2010;132:031301.
168. Jia Z, Hao C, Yan Y, Yang Y. Effects of nanoscale expanded graphite on the wear and frictional behaviors of polyimide-based composites. *Wear*. 2015;338:282–287.
169. Samyn P, De Baets P, Schoukens G, Hendrickx B. Tribological behavior of pure and graphite-filled polyimides under atmospheric conditions. *Polymer Engineering & Science*. 2003;43:1477–1487.
170. Mu LW, Feng X, Shi YJ, Wang HY, Lu XH. Friction and wear behaviors of solid lubricants/polyimide composites in liquid mediums. *Materials Science Forum*. 2010;654–656:2763–2766.
171. Huang L-J, Zhu P, Chen Z-L, Song Y-J, Wang X-D, Huang P. Tribological performances of graphite modified thermoplastic polyimide. *Materials Science and Engineering-Hangzhou-*. 2008;26:268.
172. Bolvari A, Glenn S, Janssen R, Ellis C. Wear and friction of aramid fiber and polytetrafluoroethylene filled composites. *Wear*. 1997;203:697–702.
173. Hooke C, Kukureka S, Liao P, Rao M, Chen Y. Wear and friction of nylon-glass fibre composites in non-conformal contact under combined rolling and sliding. *Wear*. 1996;197:115–122.

174. Zhao L-X, Zheng L-Y, Zhao S-G. Tribological performance of nano-Al_2O_3 reinforced polyamide 6 composites. *Materials Letters*. 2006;60:2590–2593.

175. Garcia M, De Rooij M, Winnubst L, van Zyl WE, Verweij H. Friction and wear studies on nylon-6/SiO_2 nanocomposites. *Journal of applied polymer science*. 2004;92:1855–1862.

176. Srinath G, Gnanamoorthy R. Sliding wear performance of polyamide 6–clay nanocomposites in water. *Composites Science and Technology*. 2007;67:399–405.

177. Meng H, Sui G, Xie G, Yang R. Friction and wear behavior of carbon nanotubes reinforced polyamide 6 composites under dry sliding and water lubricated condition. *Composites Science and Technology*. 2009;69:606–611.

178. Srinath G. Gnanmoothy R. Effects of organoclay addition on the two bodies wear characteristic of polyamide 6 nanocomposites. *Journal of Material Science*. 2005;40:8326–8333.

179. Zhong C, Wu Q, Guo R, Zhang H. Synthesis and luminescence properties of polymeric complexes of Cu (II), Zn (II) and Al (III) with functionalized polybenzimidazole containing 8–hydroxyquinoline side group. *Optical Materials*. 2008;30:870–875.

180. Du C-P, Li Z-K, Wen X-M, Wu J, Yu X-Q, Yang M et al. Highly diastereoselective epoxidation of cholest-5-ene derivatives catalyzed by polymer-supported manganese (III) porphyrins. *Journal of Molecular Catalysis A: Chemical*. 2004;216:7–12.

181. Moghadam M, Tangestaninejad S, Mirkhani V, Mohammadpoor-Baltork I, Kargar H. Mild and efficient oxidation of alcohols with sodium periodate catalyzed by polystyrene-bound Mn (III) porphyrin. *Bioorganic & Medicinal Chemistry*. 2005;13:2901–2905.

182. Lansdown AR. *Molybdenum Disulphide Lubrication*. Elsevier: Amsterdam, the Netherlands; 1999.

183. Cao L, Yang S, Gao W, Liu Z, Gong Y, Ma L et al. Direct laser-patterned micro-supercapacitors from paintable MoS_2 films. *Small*. 2013;9:2905–2910.

184. Sorrentino A, Altavilla C, Merola M, Senatore A, Ciambelli P, Iannace S. Nanosheets of MoS_2-oleylamine as hybrid filler for self-lubricating polymer composites: Thermal, tribological, and mechanical properties. *Polymer Composites*. 2015;36:1124–1134.

185. Yu L, Yang S, Liu W, Xue Q. An investigation of the friction and wear behaviors of polyphenylene sulfide filled with solid lubricants. *Polymer Engineering & Science*. 2000;40:1825–1832.

186. Wang J, Gu M, Songhao B, Ge S. Investigation of the influence of MoS_2 filler on the tribological properties of carbon fiber reinforced nylon 1010 composites. *Wear*. 2003;255:774–779.

4 Self-Lubricating Ceramic Matrix Composites

4.1 INTRODUCTION

Strong ionic or covalent bonds between the atoms of ceramic materials determine their strength, hardness, high melting point, modulus of elasticity (rigidity), temperature, and chemical stability. A major challenge in advanced structural ceramics is to develop long-lifetime and reproducible ceramic sliding components for use in mechanical systems. In many friction applications, high strength, corrosion resistance, and refractoriness should be combined with good tribological properties.

The combination of high wear resistance with high hardness, low density, high strength and hardness at elevated temperatures, and good corrosion resistance gives ceramic materials different applications at high temperatures and in harsh environments. Investigations of the friction and wear behavior of these materials, however, have revealed that they exhibit high friction coefficients, typically 0.5–0.8, under unlubricated sliding conditions [1–4]. Nevertheless, both the high coefficient of friction of this kind of material under dry sliding and the brittleness of ceramic–matrix itself limit its practical application in tribological areas.

To take advantage of properties of advanced ceramic materials, their friction coefficient must be reduced to 0.1 or lower. To achieve low-fraction coefficient (i.e., 5, 0, 1), using liquid lubrication is very common in most of the applications [5–7]. Therefore, use of lubricants is not effective in some applications in which the ceramic part involves in some environments such as the conditions of high temperature, vacuum, or corrosive environment. For example, the bearings and seals in advanced low-heat rejection engines and gas turbines are required to operate over a temperature range that exceeds the capabilities of conventional liquid lubricants [8–10].

Liquid lubrication allows a decrease in the coefficient of friction dramatically. In addition, deposited solid-lubricant films are effective in reduction of the friction coefficient and the wear rate; removal of the film by wear can be a concern in operating systems. This limitation can be avoided by maintaining a constant supply of solid-lubricant material at the sliding contact by embedding the solid lubricant as a second phase in the ceramic matrix to synthesize the self-lubricating ceramic matric composites. Wedeven et al. [11,12] have reported a low friction coefficient for solid-lubricated silicon nitride in a rolling contact at temperatures up to 538°C. The observed low-friction coefficient was obtained by burnishing graphite on the contacting surfaces.

Gangopadhyay et al. [6] investigated the tribological properties of alumina coated with a composite solid-lubricant coating containing silver, antimony trioxide, and barium fluoride sliding against uncoated alumina in air. At room temperature (RT), the friction coefficient of alumina without the self-lubricating ceramic composite coating was 0.40, whereas the friction coefficient was reduced to 0.11 by applying the self-lubricating ceramic composite coating. This low value of the friction coefficient was maintained up to 500°C, but the wear rate is high.

As liquid lubricants work inefficiently at a high temperature, and coatings have a limited service life, continuous supply of a solid lubricant to the ceramic surface may be provided by the ceramic material itself if it is a composite material with the ceramic matrix containing dispersed solid lubricant [13,14], but there is only limited information in the literature on the use of solids in the lubrication of ceramics. During the friction action, the solid lubricant smears over the contact surface and forms a transfer film reducing the coefficient of friction. Moreover, solid lubricants can exhibit excellent lubricating abilities in a wide range of temperatures because the lubricants can promote the formation of well-covered lubricating films on the surfaces of ceramics that can work effectively under different temperatures [15–17]. In recent years, ceramic matrix high-temperature self-lubricating composites have attracted the attention of many researchers. In addition, searching high-temperature self-lubricating composites has been driven by the demand put forth by the advanced technological systems such as high-performance gas turbine engines and aerospace applications in which the tribological components are supposed to have good tribological behaviors from RT to high temperature.

There are several potentially well-known solid lubricants for self-lubricating ceramic matrix composites which are as follows: graphite, hexagonal boron nitride (hBN), sulfides, selenides, and tellurides (chalcogenides) of metals (e.g., molybdenum disulfide), oxides (B_2O_3, MoO_2, ZnO, Re_2O_7, TiO_2, and CuO), and soft metals (bismuth, tin, silver, indium, and lead).

On the contrary, the mechanical properties of self-lubricating ceramic matrix composites are destroyed because of incorporation of the layered structural solid-lubricant phase [18,19]. In these situations, it is necessary to synthesize a high-strength and high-toughness self-lubricating ceramic composites. To improve the mechanical properties and realize self-lubrication performance of the ceramic composites, several actions have been made recently. The main methods are shown as follows: adding solid lubricants into the ceramic matrix to develop the self-lubricating ceramic composites [20–22], by *in situ* reaction method to fulfill ceramic composites self-lubricating properties [23–25], impregnating solid-lubricants [26,27], coating lubricating films [28,29], and laminating different composites to realize self-lubricating performance [30].

This chapter reviews the results of a systematic study conducted to evaluate the feasibility of achieving a low-friction coefficient in self-lubricating ceramic matrix composites. The feasibility of lubricating different ceramic matrices with intercalated several solid lubricants was investigated.

4.2 NICKEL-MATRIX COMPOSITES

4.2.1 Ni₃Al

Intermetallic compounds are promising materials for several high-temperature application resistances, electronic devices, and magnets as structural or nonstructural materials (heat resistance, corrosion conductors). Ni_3Al intermetallic compound has been extensively studied for structural applications, heat resistance, and corrosion resistance because of the low density (7.5 g/cm^3), high melting points (1668 K), high thermal conductivity, as well as high corrosion and oxidation resistance at high temperatures [31]. These excellent properties have made Ni$_3$Al as the attractive high-temperature structural material and the corrosion-resistant material for a range of engineering applications such as gas turbine hardware, high-temperature dies and molds, cutting tools, and heat-treatment fixtures [32]. It has a broad application prospect in the civil and military industrial field, especially for the field of tribology. However, polycrystalline Ni$_3$Al intermetallic compounds show brittleness at room and elevated temperatures arising from an extrinsic environmental effect [33], which severely restricts the tribological property of the material.

Several researches indicated that Ni$_3$Al may be an excellent matrix for high-temperature self-lubricating composite owing to its high-temperature strength, good-oxidation resistance, and corrosion-resistance behavior. Taking into account the excellent high-temperature performance and the potential applications in the field of tribology of Ni$_3$Al-based intermetallic alloys, several lubrication phases have been selected to prepare Ni$_3$Al self-lubricating composites. For example, the self-lubricating composites that consist of Ni$_3$Al matrix with Cr/Mo/W, Ag, and BaF$_2$/CaF$_2$ additions exhibit the low-friction coefficient and wear rate at a wide temperature range from RT to 1000°C.

To obtain high-temperature self-lubricating materials with well tribological and mechanical properties, suitable solid-lubricant selection is very important. As many ceramic composites are used in high temperature, no single materials can provide adequate lubricating properties over a wide temperature range from RT to high temperatures (800°C or even 1000°C); many efforts are made to a synergetic lubricating action of the composite lubricants, namely, the combination of low-temperature lubricant and high-temperature lubricant [34]. The conventional solid lubricants, such as MoS$_2$ and graphite, cannot meet the demand on tribological and mechanical properties due to their inadequate oxidation resistance of MoS$_2$ and graphite in air above 500°C.

hBN has been considered an effective solid lubricant for high-temperature applications as it has a graphite-like lamellar structure. For its superior adherence and thermochemical stability, hBN is an ideal solid lubricant for the temperature above 500°C [35]. However, the nonwettability and poor sinterability of hBN would restrict its applications. Except for the above layered lubricants, soft noble metal Ag and Au should be a promising lubricant for Ni$_3$Al at low temperatures (below 450°C) due to the low shear strength and stable thermochemistry. It was found that Ag added into the Ni$_3$Al-matrix composite exhibited no reactants between Ag and other additives detected after the hot-sintering process. Moreover, the composite with Ag had higher strength than those with graphite or MoS$_2$. Furthermore, during frictional process, Ag kept favorable thermal stability at low temperatures, whereas oxidation reaction could happen between Ag and other additives in the composite at high temperatures. It is noteworthy that the oxidation products such as AgMoO$_4$ are beneficial for improvement of lubricity. Mahathanabodee et al. [36] reported that the lubricating ability of hBN was inferior to that of graphite because the van der Waals force between the interlayers of hBN was stronger than that of graphite.

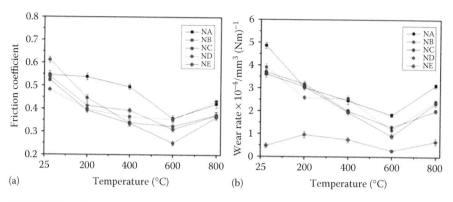

FIGURE 4.1 Variation of (a) friction coefficients and (b) wear rate of NMSC with temperatures (NA: Ni_3Al + 0% WAh + 0% TiC, NB: Ni_3Al + 10% WAh + 0% TiC, NC: Ni_3Al + 10% WAh + 0% TiC, ND: Ni_3Al + 20% WAh + 0% TiC, NE: Ni_3Al + 15% WAh + 5% TiC). (From Shi, X. et al., *Mater. Des.*, 55, 75–84, 2014.)

They found that the sintered composites containing hBN exhibited a lower friction coefficient than the based matrix.

The solid lubricant WS_2 has the lamellar structure, such as MoS_2 and graphite, and is easy to be sheared, and form transfer lubricious films between the friction pair interfaces. WS_2 has the relatively high-oxidation temperature (539°C) than MoS_2 (370°C) or graphite (325°C), so it can maintain lubricating properties at relatively higher temperatures [37].

A research investigated that the effects of different contents of solid lubricants on tribological properties of Ni_3Al matrix self-lubricating composites (NMSC), as well as the reinforced phase of titanium carbide (TiC) on the dry sliding friction-containing varying amounts of WS_2/Ag/hBN (WAh) with a weight ratio of 1:1:1, were synthesized by in situ technique using spark plasma sintering (SPS). The wear tests were conducted against Si_3N_4 ceramic ball at the load of 10 N and sliding speed of 0.234 m/s for 80 min from RT to 800°C [38]. Figure 4.1 shows the correlation between the friction coefficients and wear rates of NMSC with temperature. As shown in Figure 4.1a, embedding WAh is effective in reducing coefficient of friction (COF), and the friction coefficients of NMSC are less than Ni_3Al ceramics at all tested temperatures. In addition, the coefficient of friction gradually decreased with increasing temperature from 25°C to 600°C and then increased when the temperature increased up to 800°C for all the samples. The friction coefficients were observed to decrease from 0.61 to 0.48 at RT and from 0.43 to 0.36 at 800°C with and addition of WAh. Figure 4.1b shows the variation of wear rates of NMSC with temperatures. It could be found that the wear rate of NMSC is reduced by embedding the solid lubricants at all ranges of temperature. Moreover, the wear rates of NMSC decreased to the lowest points with increasing temperature up to 600°C and then increased at 800°C. The most prominent decrease occurred at 400°C, in which the wear rates decreased from about 2.45×10^{-4} mm³/Nm to the lowest value of 1.96×10^{-4} mm³/Nm as the addition amount of Ag increased from 0 wt.% in Ni_3Al to 5 wt.% in Ni_3Al/10% WAh, and then slightly increased to 2.55×10^{-4} mm^3/Nm as the addition amount of Ag was 6.6 wt.% in Ni$_3$Al/20% WAh. As the temperature reached 600°C, NMSC showed the lowest friction coefficients and the least wear rates. Besides, adding the TiC plays an important role to reduce the wear rate of composites due to higher hardness (7.3 GPa) than pure Ni$_3$Al (5.3 GPa) and other composites (3.8–5.7 GPa).

Study on worn surface at room temperature exhibits that the grooves on the worn surface of Ni$_3$Al were coarser, if compared with the relatively finer grooves on the worn surface of Ni$_3$Al/10%WAh. However, there were no such grooves on the worn surface of Ni$_3$Al/10%WAh, and some delamination pits were found, indicating that the main wear mechanism was surface delamination. In addition, it was clear that the obvious tribofilms and some delamination pits were observed on the worn surface of Ni$_3$Al/15%WAh/5%TiC, and the obvious deep-parallel furrows like Ni$_3$Al could not be found. The main wear mechanism was surface delamination for Ni$_3$Al/15%WAh/5%TiC. It could be found that a relatively smooth surface and dense tribolayers called *glazes* had been formed on the worn surface of Ni$_3$Al/10%WAh when compared with Ni$_3$Al. At 600°C, the worn surface of Ni$_3$Al/10%WAh was much smoother than that of Ni$_3$Al. In addition, there were a large number of debris and local compacted debris layers existing on the worn surface of Ni$_3$Al, whereas delamination pits and some white grains appeared on the surface of Ni$_3$Al/10%WAh. Ni$_3$Al/10%WAh provided the relatively continuous tribofilm, in which the energy dispersive X-ray (EDX) analysis results have shown that the lubrication phases well covered the worn surface that the presence of the glaze layer at high temperatures played an important role in tribological properties of material. For Ni$_3$Al/15%WAh/5%TiC, it could be observed that a continuous and dense tribofilm appeared on the worn surface of Ni$_3$Al/15%WAh/5%TiC. As the temperature reached 600°C, the material became much softer. Plastic deformation occurred on the worn surface of Ni$_3$Al/15%WAh/5%TiC during the sliding process. A tribofilm was formed, resulting in the low-friction coefficient and wear rate.

Inorganic salts are obvious candidates for consideration owing to low shear strength and high ductility at elevated temperatures. The high-temperature lubricious behavior of some sulfates, chromates, molybdates, and tungstates has been extensively studied for different ceramic matrices such as Ni$_3$Al [39–43]. Fluorides exhibit high-temperature solid lubricity to provide low friction coefficient and wear according to the previous studies [44,45]. Ni$_3$Al-Cr-Ag-BaF$_2$/CaF$_2$ composites were synthesized by powder metallurgy technique [46]. X-ray diffraction (XRD) results indicated that components in the sintered Ni$_3$Al-matrix composites did not react on each other, and no new compound was formed during the fabrication process. XRD patterns of worn surfaces after frictional tests presented that at 600°C, BaCO$_3$ in the form of weak peak appears, and at 800°C, no BaF$_2$ peaks are present, but BaCrO$_4$ peaks were found. Fluorides served as high-temperature lubricants and exhibited a good reduce-friction performance at 400°C and 600°C. However, at 800°C, BaCrO$_4$ formed on the worn surface due to the tribochemical reaction at high temperatures provided an excellent lubricating property.

Ti$_3$SiC$_2$ has been proved to be promising tribological material as an additive to the self-lubricating composites [47–49]. MoS$_2$ as solid lubricant has been extensively studied [50]. The results show that MoS$_2$ can work well at low temperatures, especially in vacuum, but it can be easily oxidized at high temperatures.

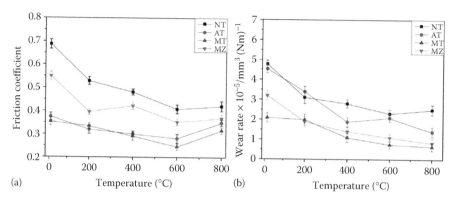

FIGURE 4.2   Variations of (a) friction coefficients and (b) wear rates of samples. (From Yao, J. et al., *J. Mater. Engi. Perform.*, 24, 280–295, 2015.)

Hence, MoS_2 is often selected as solid lubricant at low temperatures. The Ni_3Al (NT), $Ni_3Al–Ag–Ti_3SiC_2$ (AT), $Ni_3Al–MoS_2–Ti_3SiC_2$ (MT), and $Ni_3Al–MoS_2–ZnO$ (MZ) composites are fabricated using SPS [51]. The wear and lubricating mechanisms of NMSCs from 25°C to 800°C were investigated. The composite powders of Ni_3Al matrix consist of commercially available Ni, Al, Cr, Mo, Zr, and B powders (30–50 μm in average size, 99.9 wt.% purity) by atomic ratio of 4.5Ni:1Al:0.333Cr:0.243Mo:0.0047Zr:0.0015B. The fabrication process of Ti_3SiC_2 powder (5 μm in average size, 95.0 wt.% purity). The average grain sizes of Ag, ZnO, and MoS_2 powders were about 20–40 μm. Figure 4.2a shows the correlation in friction coefficients of NT, AT, MT, and MZ against Si_3N_4 balls at different temperatures under an applied load of 10 N. It can be seen that the friction coefficient of NT at RT is relatively high (0.69). As the temperature increases from RT to 800°C, the friction coefficient of NT decreases slightly from 0.69 to 0.45. The friction coefficients of MZ are smaller than that of NT throughout the test temperatures and in the range of 0.38–0.55. AT and MT have lower friction coefficient compared with NT and MZ. In addition, the friction coefficients of AT and MT exhibit downward trend from 200°C to 600°C, whereas the upward trend for all samples is observed when the temperature is raised above 600°C. The fluctuations of friction coefficients, as can be seen from the error bar, are higher for the NT and AT, if compared with that of the MT and MZ. Figure 4.2b shows the variation of wear rates of samples at different temperatures. For NT, the wear rate reaches its minimum value of 2.69105 mm^3/Nm at 600°C and increases as the temperature increases to 800°C. It can be found that the wear rates of MT and MZ decrease sharply from RT to 400°C and continue to decline slowly up to 800°C. As the temperature increases from RT to 800°C, the wear rates of AT decrease from 4.5 to 1.89105 mm^3/Nm, except for the wear rate at 600°C, which is higher than that at 400°C and 800°C. The reason for the relatively high wear rate at 600°C is investigated in discussion in detail. As the temperature increases to 200°C, the MT and MZ show a comparable wear rate of about 2.09105 mm^3/Nm. The wear rates of MT are lower from 400°C to 800°C, if compared with that of the MZ. At 800°C, the MT exhibits excellent wear resistance, and the wear rate is about 0.89105 mm³/Nm. The fluctuations of wear rates, as can be seen from the error bar, are higher for the NT and AT, if compared with that of the MT and MZ.

The wear tracks on the worn surfaces of AT surprisingly show the parallel ridges and deep grooves on the worn surface at the lower temperatures, whereas the coarse grooves and plastic flow at the higher temperatures. This is to be expected that, with the soft lubricant of Ag addition. In other words, the tribological property of MZ has been improved by adding MoS_2 and ZnO lubricants. This result can be rationalized with the help of the wear mechanisms operating in MZ during dry sliding, such as plastic deformation, oxidation, and abrasive wear. Among them, the plastic deformation is the dominant wear mechanism. Obviously, the MT shows the excellent synergetic lubricating action. Morphologies of worn surfaces tested from RT to 600°C show plastic deformation with relatively smooth surface, which confirm the very low friction coefficients and wear rates. Moreover, the morphology of worn surface tested at 800°C shows abrasive wear and oxidation wear together with scratches and debris particles, leading to the increase in friction coefficient. To clarify the microstructure and the formation mechanism of the friction layer of MT, the subsurface analysis (at RT, 600°C, and 800°C) are carried out on the worn surface by cross-sectioning it perpendicular to the sliding direction. Figure 4.3 shows the wear track morphology of Si_3N_4 ball against MT at 600°C. As shown in Figure 4.3, the worn surface of Si_3N_4 ball is partly covered with smooth lubricating films, which also exists on worn surface of MT.

As a high-temperature solid lubricant, and similar to $CaWO_4$ and $CaMoO_4$, $BaMoO_4$ has scheelite structure and adequate thermophysical properties [52,53]. However, the lubricious behavior of $BaMoO_4$ has not been explored in detail. Recently, $BaCrO_4$ has attracted much attention due to its lubricating property at a wide temperature range [39]. $BaCrO_4$ has an orthorhombic structure, and its thermal data show that the $BaCrO_4$ phase is thermally stable to 850°C [54,55]. Therefore, they could be expected as promising high-temperature solid lubricants for Ni_3Al.

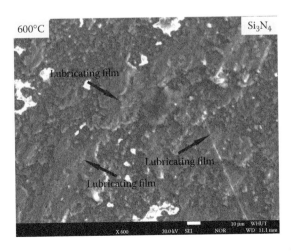

FIGURE 4.3   SEM image of worn surface of Si_3N_4 ball against MT at 600°C. (From Yao, J. et al., *J. Mater. Engi. Perform.*, 24, 280–295, 2015.)

4.2.2 NiAl

NiAl matrix compound has attractive properties such as high melting point (1638°C), high thermal conductivity, low density (5.86 g/cm^3), high Young's modulus (294 GPa), large thermal conductivity (76 W/mK), and excellent oxidation resistance at high temperatures (~1000°C) [56–58]. However, NiAl is not widely used in structural applications due to its poor ductility at ambient temperatures and low strength and creep resistance at elevated temperatures. On one hand, some researches have been carried out to study the tribological behaviors of NiAl at RT, indicating that the NiAl possesses good tribological properties [59,60]. NiAl is selected to be used as candidate for structural material because of the combination properties, especially the attractive tribological performance [59]. Over the past years, some studies have been carried out to investigate the tribological behaviors of NiAl, which have provided useful information regarding application of NiAl matrix in sliding wear situations, indicating that the NiAl possesses good tribological properties [60,61].

Wear testing of three NiAl alloys containing 45 at.%, 48 at.%, 50 at.% aluminum indicated that all of them had low friction coefficient (0.25–0.35) and wear rate (1.5–2.4 × 10^{-5} mm^3/Nm) at RT. On the other hand, to further improve the wear resistance of NiAl intermetallic materials at high temperatures, many efforts have been made in recent years. NiAl–31BaF$_2$–19CaF$_2$ (mass %) exhibited the low friction coefficients (0.3–0.4) and wear rates (3–4 × 10^{-5} mm^3/Nm) [40]. Zhu et al. [62] selected ZnO and CuO as lubricants in NiAl matrix composites due to their high melting point and favorable wear resistance at high temperatures. It was found that NiAl matrix composites with addition of ZnO showed the lowest wear rate (7 × 10^{-6} mm^3/Nm) at 1000°C, whereas CuO addition into NiAl matrix composites exhibited the self-lubricating performance and the best tribological properties at 800°C. From the tribology point of view, the addition of effective solid lubricants becomes an excellent solution to promote the application of NiAl matrix materials at different temperatures. In a study [63], it is reported that Ti$_3$SiC$_2$ as solid lubricant in NiAl alloys can promote the strength and improve tribological performances of NiAl matrix self-lubricating composites effectively at high temperatures.

Ozdemir et al. [64] have researched the friction coefficient and wear rate of NiAl intermetallic compound under different loads, which was produced by pressure-assisted combustion synthesis. The friction coefficient for NiAl matrix compound under the load of 2 N was 0.73, whereas it was 0.53 under the load of 10 N. Moreover, the friction coefficient presented the tendency decrease with the increase in load. Moreover, NiAl intermetallic materials showed good tribological properties with the addition of soft oxide at high temperatures [65–67].

Alloying is one of the effective approaches that has been used successfully to improve the RT fracture toughness, yield strength, and ductility of brittle intermetallics. NiAl–28Cr–6Mo eutectic alloys are regarded as the most logical choice of the multi-element system examined to date because of their relatively high melting point, good thermal conductivity, and high elevated temperature creep resistance as well as higher fracture toughness [68,69]. Thus, NiAl–28Cr–6Mo alloy may be an excellent matrix for high-temperature self-lubricating composite. Recently, NiAl matrix high-temperature self-lubricating composites also have been explored [62,70]. NiAl matrix composite with various high-temperature solid lubricants, such as oxide and fluoride, provide excellent lubricating properties at elevated temperatures. It is well known that the addition of soft oxide is one of the effective approaches to reduce friction and wear at high temperatures because the softening oxide could offer low shear strength and high ductility and the formation of a glaze film would protect the sliding surface from heavy wear. NiAl, NiAl–Cr–Mo alloy, and NiAl matrix composites with addition of oxides (ZnO/CuO) were fabricated by powder metallurgy route [62]. It was found that some new phases (such as $NiZn_3$, $Cu_{0.81}Ni_{0.19}$, and Al_2O_3) were formed during the fabrication process due to a high-temperature solid-state reaction. The results indicated that the monolithic NiAl had high-friction coefficient and wear rate at elevated temperatures due to poor mechanical properties. The incorporation of Cr(Mo) not only enhanced mechanical properties evidently but also improved high-temperature tribological properties greatly. NiAl matrix composite with addition of ZnO showed superior wear resistance at 1000°C among the sintered materials, which was due to the formation of the ZnO layer on the worn surface. NiAl matrix composite with addition of CuO exhibited self-lubricating performance at 800°C, which was attributed to the presence of the glaze layer containing CuO and MoO_3. Meanwhile, it had the best tribological properties among the sintered materials at 800°C.

In addition, CaF_2 added into NiAl matrix composite exhibited favorable friction coefficient about 0.2 and excellent wear resistance about 1×10^{-5} mm³/Nm at high temperatures (800°C and 1000°C) [70]. The excellent self-lubricating performance was attributed to the formation of the glaze film on the worn surface, which was mainly composed of $CaCrO_4$ and $CaMoO_4$ as high-temperature solid lubricants. However, the composite had poor tribological performance at low temperatures. Addition of Ag evidently reduced friction coefficient and enhanced wear resistance at low temperatures. It indicated that Ag functioned as a favorable solid lubricant for NiAl intermetallic at low temperatures. However, it was adverse to friction and wear at elevated temperatures because of the decrease in the strength of material. On the whole, NiAl–Cr–Mo–CaF_2–Ag composite provided self-lubricating properties at a broad temperature range between RT and 1000°C (Figure 4.4). Especially at 800°C,

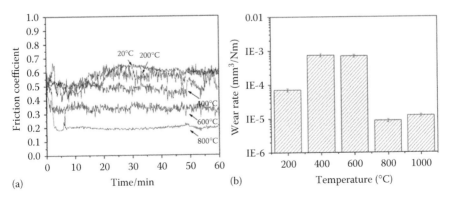

FIGURE 4.4 Variations of (a) friction coefficients and (b) wear rates of NiAl–Cr–Mo–CaF_2–Ag at different temperatures (tested at an applied load of 10 N and sliding speed of 0.2 m/s against Si_3N_4 ceramic ball).

the composite offered excellent friction reduction about 0.2 and wear resistance about 7×10^{-5} mm^3/Nm at high temperatures. The low-friction coefficient at a wide temperature range could be attributed to the synergistic effect of Ag, CaF$_2$, CaCrO4, and CaMoO$_4$.

Some conventional solid lubricants like MoS$_2$ and WS$_2$ can work well at low temperatures because of their special structure [50,71–74]. The maximum useful temperature for solid lubricant of MoS$_2$ is limited to about 400°C in air, whereas they can be easily oxidized at high temperatures [71,72]. Ti$_3$SiC$_2$ ceramics have been proved to be the promising tribological materials at high temperatures [75,76]. As a new solid self-lubricating material, Ti$_3$SiC$_2$ has been proved to be the promising tribological material at high temperatures with low-friction coefficient and wear rate [49,75,77]. Meager information is available as regards the development of NiAl matrix self-lubricating composites by the use of Ti$_3$SiC$_2$ and MoS$_2$ lubricating phases and TiC-enhanced phase. The present investigation is aimed at preparing NiAl matrix self-lubricating composites containing a fixed amount of Ti$_3$SiC$_2$ and varied amounts of MoS$_2$ by SPS. The possibility of synergetic action of Ti$_3$SiC$_2$ and MoS$_2$ lubricants has been explored by carrying out dry sliding wear from RT to 800°C. The synergetic lubricating effect of Ti$_3$SiC$_2$ and MoS$_2$ lubricants on the friction and wear behavior of the composites has been analyzed and discussed [78]. NiAl-based alloy (NA) and NiAl matrix self-lubricating composites with 5 wt.% Ti$_3$SiC$_2$ and different contents (0, 3, 5, and 7 wt.%) of MoS$_2$ (NAT, NATM3, NATM5, and NATM7) were prepared by SPS. The composite powders of NiAl matrix were composed of commercially available Ni, Al, Mo, Nb, and B powders (30–50 μm in average size, 99.9 wt.% in purity) by an atomic ratio of 48:50:1:1:0.02.

The variations of friction coefficients of NA, NAT, and NATMs with temperatures were given in Figure 4.5a. It could be found that the friction coefficient of NA sharply increased with the increase in temperature and obtained the largest value of about 0.71 at 800°C. With the addition of Ti$_3$SiC$_2$, the friction coefficient of NAT initially increased at 200°C while gradually decreased with the increase in temperature and obtained the lowest value at 800°C. With the addition of Ti$_3$SiC$_2$ and MoS$_2$ lubricants,

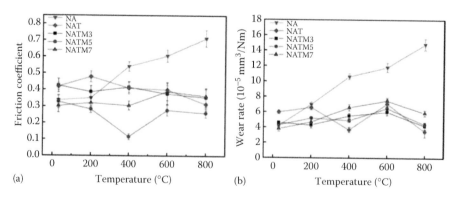

(a)   (b)

FIGURE 4.5   The variations of friction coefficients (a) and wear rates (b) of NA, NAT, NATM3, NATM5, and NATM7 with temperatures. (From Shi, X. et al., *Mater. Des.*, 55, 93–103, 2014.)

NATMs when compared with NA and NAT exhibited lower friction coefficients over a wide temperature range. Below 400°C, for the addition of MoS$_2$, the friction coefficients of NATMs when compared with NAT dropped to 0.13–0.42, indicating that the MoS$_2$ enhanced the tribological performance at low temperatures. With the increasing of temperature, the friction coefficients of NATMs were much lower than that of NA when the temperature was in the range of 400°C–800°C. Moreover, it could be found that the NATM5 had the lowest value (under 0.30) at 200°C–800°C in all samples. Figure 4.5b shows the variations of wear rates of NA, NAT, and NATMs with temperatures. It could be seen that the wear rates of NA gradually increased from 4.30×10^{-5} mm^3/Nm to 1.45×10^{-4} mm^3/Nm with the increase in temperature. The wear rates of NAT exhibited large fluctuation from RT to 800°C. It got to the lowest value of about 3.8×10^{-5} mm^3/Nm at 400°C. With the addition of Ti$_3$SiC$_2$ and MoS$_2$ lubricants, the wear rates of NATMs were much lower than that of NA and exhibited the similar variation tendency when the temperature was above 200°C. Moreover, it could be found that the wear rates of NATM5 had the most stable value at all the tested temperatures. NATM5 exhibited excellent tribological performance among all the samples over a wide temperature range. According to the aforementioned friction and wear results, it could be concluded that NATMs when compared with NA and NAT exhibited lower friction coefficients and wear rates over a wide temperature range. The results indicated that the addition of MoS$_2$ was responsible for the lower friction coefficients of NATMs between RT and 400°C [79], as well as the Ti$_3$SiC$_2$ enhanced the self-lubrication properties at high temperatures [75,77]. Consequently, it was inferred that the addition of Ti$_3$SiC$_2$ and MoS$_2$ lubricants was an effective way to widen operating temperature range of NiAl matrix composites.

It could be seen that the morphologies of worn surfaces of NA were different with the changing of temperature. As shown in Figure 4.6a, there were obvious cracks and pits on the worn surface of NA at RT, indicating that the main wear mechanism was microfracture. At 200°C and 400°C, as shown in Figure 4.6b and c, the worn surfaces of NA became flat, and the continuous and homogenous scratches could be seen. When the temperature increased up to 600°C, a loose tribofilm formed during the sliding process because abundant wear particles adhered on the worn surface as shown in Figure 4.6d. At 800°C, the worn surface of NA was covered by a tribofilm which was deeply torn as shown in Figure 4.6e. Serious delamination could also be observed on the deformed surface. The main wear mechanism was dominant surface deformation. It could be seen that the worn surfaces of NAT at different temperatures were much smoother than those of NA, and obvious tribofilms could be observed on the worn surfaces of NAT. Parallel furrows and spallations were shown in Figure 4.7a–c, indicating that the main wear mechanisms were plowing and delamination at low temperatures. As the temperature increased up to 600°C, some thin and discontinuous tribofilms were formed in some areas. Most of the worn surfaces were covered by wear particles, as shown in Figure 4.7d. The tribofilms of NAT when compared with NA became much denser at 600°C. It was attributed to the reinforcement of TiC particles during the sliding process. EDX analysis indicated that abundant amounts of O, Al, and Ni, as well as a small amount of Ti, Si, and C, were also observed on the worn surface of NAT. Abundant oxides of aluminum and nickel were formed during the sliding process. With a further increase in temperature, the worn

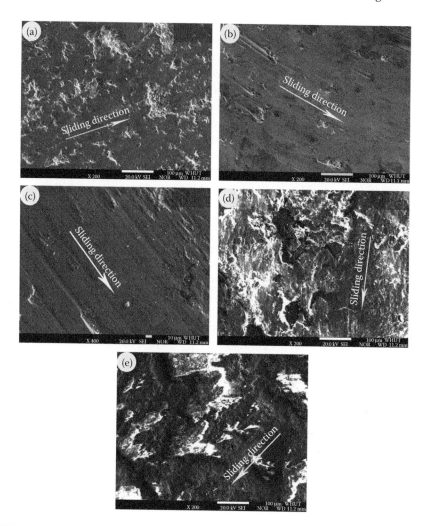

FIGURE 4.6 Typical electron probe morphologies of worn surfaces of NA at different temperatures: (a) RT, (b) 200°C, (c) 400°C, (d) 600°C, and (e) 800°C. (From Shi, X. et al., *Mater. Des.*, 55, 93–103, 2014.)

surface became much smoother as shown in Figure 4.7e. Moreover, the continuous tribofilms and mild scratches were formed on the worn surface of NAT. The presence of a compacted transfer layer of wear debris reduced Si_3N_4 ceramic ball-NAT direct contact and provided the low shear strength junctions at the interface, as well as decreased the energy required to shear these junctions; hence, the friction coefficient was reduced. It could be observed that the worn surfaces of NATM5 when compared with NA at each temperature turned much smoother. At RT, as shown in Figure 4.8a, the abundant lamellar particles adhered to the worn surface of NATM5, indicating that the main wear mechanism was adhesive wear. As the temperature increased to 200°C, a few tiny cracks occurred on the worn surface (Figure 4.8b). At 400°C, it

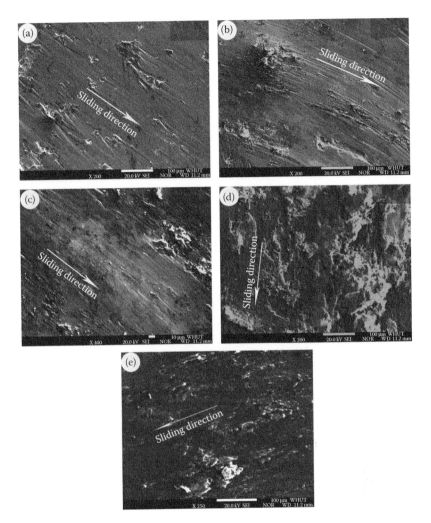

FIGURE 4.7 Typical electron probe morphologies of worn surfaces of NAT at different temperatures: (a) RT, (b) 200°C, (c) 400°C, (d) 600°C, and (e) 800°C. (From Shi, X. et al., *Mater. Des.*, 55, 93–103, 2014.)

could be seen from Figure 4.8c that a thin tribofilm was formed on the worn surface, resulting in the lowest friction coefficient and wear rate. As the temperature increased to 600°C, a dense and complete tribofilm was present on the worn surface of NATM5 as shown in Figure 4.8d. However, the friction coefficient and wear rate increased at 600°C, The thermal stress-induced tribochemical oxidation to generate metal oxide on the friction surface, and abundant MoO_3 was formed on the worn surface during the sliding process as the temperature increased [80]. All the above analyses indicated that addition of MoS_2 and Ti_3SiC_2 into the NiAl based self-lubricating composites was a direct and effective way to widen operating temperature range. MoS_2 played a self-lubricating role at low and medium temperatures, whereas the Ti_3SiC_2 was effective

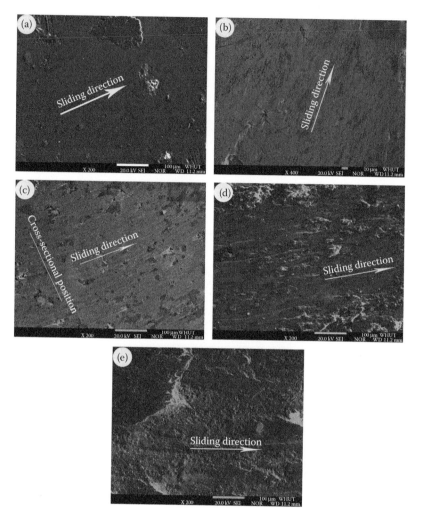

FIGURE 4.8 Typical electron probe morphologies of worn surfaces of NATM5 at different temperatures: (a) RT, (b) 200°C, (c) 400°C, (d) 600°C, and (e) 800°C. (From Shi, X. et al., *Mater. Des.*, 55, 93–103, 2014.)

at high temperatures. The synergetic lubricating action of MoS_2 and Ti_3SiC_2 was well realized in NATM5. Meanwhile, the presence of TiO_2, SiO_2, and MoO_3, on the worn surface of NATM5, is confirmed by the XPS spectra.

Another research explored the concept of synergetic lubricating action of MoS_2 with Ti_3SiC_2, as well as WS_2 with Ti_3SiC_2 from RT to 800°C [81]. The NiAl-based self-lubricating composites were prepared through the SPS, in which MoS_2, WS_2, and Ti_3SiC_2 were used as lubricants to improve the tribological properties of the composites. NA without lubricant as well as NiAl containing PbO have also been prepared and tested. Taking into account the tribological properties at low temperatures and high temperatures, each lubricant (MoS_2, Ti_3SiC_2, and WS_2) in the binary

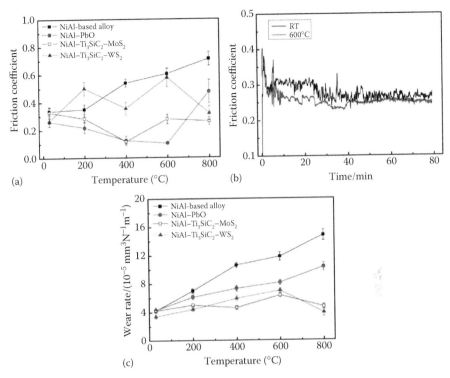

FIGURE 4.9 The variation of friction coefficients (a), representative variation of friction coefficients of NiAl–Ti$_3$SiC$_2$–MoS$_2$ with time (b), and wear rates (c) of NiAl-based self-lubricating composites at different temperatures. (From Shi, X. et al., *Wear*, 310, 1–11, 2014.)

lubricant (Ti$_3$SiC$_2$–MoS$_2$, Ti$_3$SiC$_2$–WS$_2$) is selected 5 wt.%. Compared with each single lubricant, the PbO lubricant is also selected 5 wt.%. Hence, the weight fractions of lubricant in NiAl-based self-lubricating composites were fixed at 5 wt.%PbO, 5 wt.%Ti$_3$SiC$_2$–5 wt.%MoS$_2$, 5 wt.%Ti$_3$SiC$_2$–5 wt.%WS$_2$, respectively.

The variation of friction coefficients of NiAl-based self-lubricating composites at different temperatures was given in Figure 4.9a. It could be found that the friction coefficients of NA sharply increased with the increase in temperature and obtained the largest value of about 0.70 at 800°C. With the addition of PbO, the friction coefficients of NiAl–PbO obviously decreased with the increase in temperature and obtained the lowest value of about 0.10 at 600°C. However, the friction coefficient of NiAl–PbO increased up to 0.45 at 800°C, demonstrating that the application of PbO was in a restricted temperature range. Hence, the NiAl–PbO possessed self-lubricating performance below 600°C. With the addition of binary lubricant Ti$_3$SiC$_2$–MoS$_2$, NiAl–Ti$_3$SiC$_2$–MoS$_2$ exhibited lower friction coefficients (below 0.30) from RT to 800°C, indicating that the binary lubricant of Ti$_3$SiC$_2$–MoS$_2$ could play an excellent self-lubricating role over a wide temperature range. With the addition of Ti$_3$SiC$_2$–WS$_2$, the friction coefficients of NiAl–Ti$_3$SiC$_2$–WS$_2$ did not show obvious variation trend with the increase in temperature. It seemed that the friction coefficient increased sharply from RT to 200°C, and then it decreased quickly at higher temperature of 400°C. However, the friction coefficient increased at 600°C again and then decreased at 800°C. Figure 4.9b shows the variation of wear rates of NiAl-based self-lubricating composites at different temperatures. It could be seen that the wear rates of NA gradually increased from 4.30×10^{-5} mm³/Nm to 1.45×10^{-4} mm³/Nm with the increase in temperature, whereas the wear rates of NiAl–PbO when compared with NA showed the similar variation tendency. With the addition of Ti_3SiC_2–MoS_2, NiAl–Ti_3SiC_2–MoS_2 when compared with NA shows lower wear rates, and the wear rates were under 6×10^{-5} mm³/Nm over a wide temperature range. The wear rates of NiAl–Ti_3SiC_2–WS_2 increased from RT to 600°C and then decreased with the increase in temperature. According to the above friction and wear results, it could be concluded that NiAl–Ti_3SiC_2–MoS_2 exhibited the lower friction coefficients and wear rates over a wide temperature range. Compared with the NA, after adding Ti_3SiC_2–MoS_2, the friction coefficient of NiAl–Ti_3SiC_2–MoS_2 significantly decreased and kept a consistent value below 0.30, whereas the wear rates were substantially reduced from 1.2×10^{-4} mm³/Nm to 4.7×10^{-5} mm³/Nm at 600°C. Especially, at 400°C and 800°C, NiAl–Ti_3SiC_2–MoS_2 showed the excellent self-lubricating performance. Hence, it was inferred that binary lubricant Ti_3SiC_2–MoS_2 was effective to widen operating temperature range of NiAl-based composites. The results indicated that the wide temperature-range self-lubrication properties were realized by MoS_2 at low and medium temperatures [79], as well as Ti_3SiC_2 at high temperatures [75,76]. With the addition of PbO or binary lubricant Ti_3SiC_2–WS_2, the self-lubrication properties of the NiAl-based composites were unstable over a wide temperature range.

From the morphologies of worn surfaces of NA, there were obvious signs of cracks and pits on the worn surface of NA at RT, which could mean that the main wear mechanism was microfracture. At 200°C and 400°C, the worn surfaces of NA became flat, whereas the continuous and homogenous scratches could be seen. As the temperature increased up to 600°C, a loose tribofilm formed during the sliding process adhered on the worn surface due to abundant wear particles. At 800°C, the worn surface of NA was covered by a tribofilm that was deeply torn. Serious delamination could also be observed on the deformed surface. The main wear mechanism was dominant surface deformation, indicating that the plasticity of NA under dry sliding process required a critical contact pressure. Therefore, it was to be understood that the combined effect of high contact pressure and high temperature resulted in observed delamination along with deformed surface. On the worn surface of NiAl-PbO, it was obvious that abundant pits were found on the worn surface. Adhesive wear was the main wear mechanism of NiAl–PbO at RT. At 200°C, it was apparent that there was the evidence of a significant amount of deep grooves as well as pits. As the temperature increased to 400°C, the grooves on the worn surface of NiAl–PbO became deeper and coarser, if compared with the relatively finer grooves of worn surfaces at 200°C. Moreover, the corresponding wear mechanisms transferred from microcutting at 200°C to microplowing and plastic deformation. At 600°C, a gray thick tribofilm was formed on the worn surface of NiAl–PbO. When the temperature increased to 800°C, the worn surface of NiAl–PbO became rough and serious delamination could also be observed on the deformed surface. The EDX analysis of the tribofilm showed that the gray tribofilm consisted of lead oxides.

The lead oxides were squeezed out and spread on the worn surface during the sliding process. The tribofilm could contribute to the low friction coefficient of NiAl–PbO at 600°C. It played an important role to protect the material. It could be observed that the worn surface of NiAl–Ti$_3$SiC$_2$–MoS$_2$ when compared with NA at different temperatures turned much smoother. At RT, there were abundant lamellar particles that adhered to the worn surface of NiAl–Ti$_3$SiC$_2$–MoS$_2$, meaning that the main wear mechanism was adhesive wear. During the sliding process, the wear debris that was formed at earlier stage moved between the two surfaces, and the major wear mechanism changed from two-body abrasion to three-body abrasion. As the temperature increased to 200°C, plenty of tiny cracks (highlighted by the white oval) occurred on the worn surface. At 400°C, it could be seen that a thin tribofilm was formed on the worn surface, which could be contributed to the lowest friction coefficient and wear rate. Moreover, scratches as well as plowing grooves could be observed on the worn surface of NiAl–Ti$_3$SiC$_2$–MoS$_2$. The TiC particle in NiAl–Ti$_3$SiC$_2$–MoS$_2$ played an important role in increasing the hardness and resisting plastic deformation. As the temperature increased to 600°C, a dense and complete tribofilm was present on the worn surface of NiAl–Ti$_3$SiC$_2$–MoS$_2$. Meanwhile, the presence of MoO$_3$ on the worn surface at 600°C was confirmed by the XPS spectra. The morphologies of NiAl–Ti$_3$SiC$_2$–WS$_2$ were similar to that of NiAl–Ti$_3$SiC$_2$–MoS$_2$ at different temperatures. At RT, obvious scratches and pits could be found on the worn surface of NiAl–Ti$_3$SiC$_2$–WS$_2$, as well as some particles adhered to the worn surface. As the temperature increased to 200°C, more pits occurred on the worn surface of NiAl–Ti$_3$SiC$_2$–WS$_2$. The worn surface turned much coarser when the temperature increased to 400°C. Obvious grooves and pits could be seen on the worn surface of NiAl–Ti$_3$SiC$_2$–WS$_2$, indicating that the main wear mechanism was microplowing. At 600°C, it could be seen that a compact and uniform tribofilm was formed on the worn surface. As the temperature increased to 800°C, a destroyed tribofilm appeared on the worn surface. The EDX analysis revealed that the content of O was 13.66 at.%, implying that the oxidation phenomenon was not serious with the addition of binary lubricant Ti$_3$SiC$_2$–WS$_2$ during the sliding process. As the temperature increased to 800°C, a destroyed tribofilm appeared on the worn surface.

Figure 4.10 was a schematic illustration that showed the wear mechanisms of NiAl–Ti$_3$SiC$_2$–MoS$_2$ during the sliding progress at different temperatures. At RT, as shown in Figure 4.10a, for the thermal stress, the low temperature (from RT to 400°C) solid lubricant (MoS$_2$) were squeezed out and spread on the worn surface during the sliding process. However, the effect of the thermal stress was restricted because of the low environment temperature. Hence, the content of lubricant on the surface was few. The worn surface was not completely covered by the lubricants, which was attributed to the abundant cracks. With the increase in temperature, the effect of the thermal stress became obvious; hence, more and more low temperature solid lubricant MoS$_2$ was squeezed out from the composites. Meanwhile, MoS$_2$ was evenly spread on the worn surface during the sliding process. Hence, a complete tribofilm containing low-temperature solid lubricant was formed on the worn surface (Figure 4.10b), and it effectively isolated the friction between the sample and Si$_3$N$_4$ counterpart ball. The friction coefficient and wear rate were decreased owing to the tribofilm. Moreover, more serious oxidation phenomenon took place

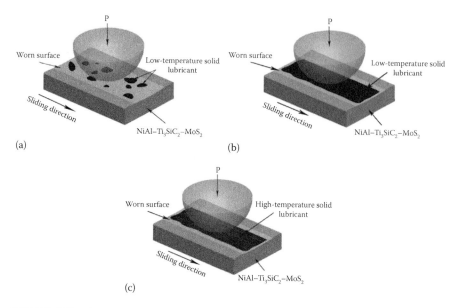

FIGURE 4.10 A schematic illustration that showed the wear mechanisms of $NiAl–Ti_3SiC_2–MoS_2$ during the sliding progress at different temperatures: (a) RT, (b) RT-400°C, and (c) 400°C–800°C. (From Shi, X. et al., *Wear*, 310, 1–11, 2014.)

with the increase in temperature during the sliding process, and the tribofilm containing low temperature solid lubricant MoS_2 was oxidized when the temperature was above 400°C. The worn surface was covered by a tribofilm containing abundant Mo–Ti–Si-Oxides (Figure 4.10c). Meanwhile, another effective tribofilm containing the high-temperature solid lubricant (Ti_3SiC_2) was formed and played a role in protecting the worn surface. According to the above analysis, we could get a conclusion that the low-temperature solid lubricant (MoS_2) was squeezed out from the composites and formed a tribofilm on the worn surface below 400°C. It effectively protected the samples at low temperatures and was beneficial to the lower friction coefficient and wear rate of the sample. With the increase in temperature, the tribofilms containing low temperature solid lubricant were gradually oxidized. Meanwhile, a new tribofilm containing high-temperature solid lubricant (Ti_3SiC_2) was formed to continuously protect the worn surface at high temperatures. The wide temperature-range self-lubrication properties were realized by low-temperature solid lubricant (MoS_2) at low and medium temperatures, as well as high-temperature solid lubricant (Ti_3SiC_2) at high temperatures. Hence, adding MoS_2 and Ti_3SiC_2 into the NiAl-based self-lubricating composites was a direct and effective way to widen operating temperature range.

Graphene, which is used as solid lubricating additive, has always been widely researched since its discovery. The distinctively excellent mechanical properties [82] such as elastic modulus (0.5–1 TPa) and tensile strength (130 GPa) has ensured its bright prospect as a reinforcement for composite materials for structural applications [83,84]. A research studied the lubricating action and different friction modalities of NiAl

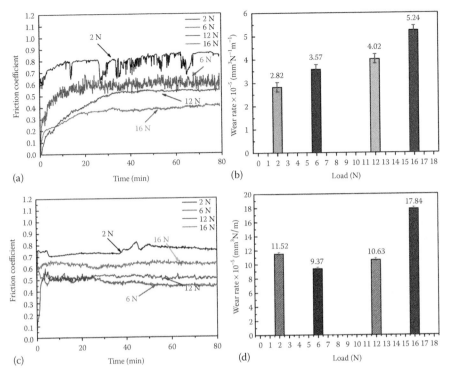

FIGURE 4.11 Friction coefficients and wear rates of NSMG (a, b) and NiAl-based alloy (c, d) sliding against Si_3N_4 ball under different loads.

self-lubricating matrix composite with graphene (NSMG) to ascertain the suitable load condition for excellent tribological performance [85]. The tribological behaviors of NSMG are investigated by sliding against a 6-mm-diameter Si_3N_4 ball with the sliding speed of 0.2 m/s under the different loads of 2, 6, 12, and 16 N for 80 min at RT. Figure 4.11a shows the typical measuring curves of the dynamic friction coefficients. Figure 4.11b exhibits the wear rates of the sample at loads from 2 to 16 N. At 2 N, NSMG has relatively high value of the friction coefficient (0.6–0.8) and low wear rate (2.82×10^{-5} mm³/Nm). Obviously, the friction coefficient curve is uneven and the value of vertical axis severely fluctuates, indicating that the contact pair runs unsteadily. Numerical fluctuation in large scope proves that the contact is interrupted. The two parts of the contact pair are transitorily separated during this process. At 6 N, The friction coefficient decreases (0.5–0.6) with increasing load up to the value of 6 N, whereas the wear rate increases to 3.57×10^{-5} mm³/Nm. At the contact load of 6 N, the friction curve is relatively smooth and steady to the curve of 2 N; the range of friction coefficient on the vertical axis is much lower than the curve at the load of 2 N. Moreover, the value of friction coefficient is smaller than that of 2 N, indicating that the friction process becomes more stable with increasing the contact load to 6 N. At 12 N, NSMG has the low values of friction coefficient (about 0.5) and wear rate (4.02×10^{-5} mm³/Nm). According to the curve, the friction

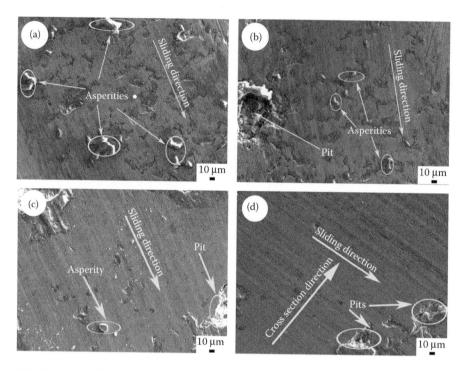

FIGURE 4.12 EPMA morphologies of worn surfaces of NSMG after tests under different contact loads: (a) 4 N, (b) 8 N, (c) 12 N, and (d) 16 N.

coefficient is slightly lower, and the friction coefficient curve is much smoother than the previous one (6 N), indicating that the friction state becomes steady with increasing the load to 12 N. At 16 N, NSMG has the values of lowest friction coefficient (0.42) and highest wear rate (5.24 × 10⁻⁵ mm³/Nm). The friction curve at the contact load of 16 N is smoother, if compared with those of 2 and 6 N. The average value of the friction coefficient of 16 N significantly decreases and is the smallest one. It can be concluded that by increasing the load from 2 to 16 N, the friction coefficient decreases and the friction curve becomes smoother.

Figure 4.12 exhibits the typical EPMA morphologies of worn surfaces of NSMG after tests. The four pictures of Figure 4.12 represent the four EPMA morphologies at the loads from 2 to 16 N. As shown in Figure 4.12, some parallel grooves are found on the worn surfaces in the four pictures, but the grooves in the pictures of load 2 N (Figure 4.12a) as well as 6 N (Figure 4.12b) were occasionally discontinuous. The discontinuity of scratches on the worn surfaces could be caused by the separation of the contact and reattachment of the ball and disk during the friction process. Moreover, there are some seriously abrupt flaws and asperities existing on the worn surfaces of NSMG at the contact loads of 2 and 6 N. As the load increases to 12 N, as shown in Figure 4.12c, the worn surface becomes flat and much smoother, and the grooves become fine, which are totally different from the narrow and deep grooves at the contact load of 2 and 6 N. Moreover, with the heavier normal load, the worn surface exhibits continuous. Meanwhile, a few corroded pits and wear debris are found on the worn surface as well. It is believed that with the increasing of the normal load, the contact stress is higher and contact areas become larger, which means the greater tendency to generate tribofilm for the effect of plastic deformation. When the load increases to 16 N, the wear scar becomes wider, and the feature of the whole worn surface is more delicate and flatter, unlike previous images at lower loads with more pits and asperities. Evidently, relatively serious plastic deformation occurs at such higher load, which will crush the asperities of worn surface to form some tiny particles. The effect of plastic deformation on pits as well as reduction of asperities and flaws on the worn surface would cut down the contribution rate of unstable contact to the friction coefficient; hence, the friction coefficient decreases at the load of 16 N. But the larger pits are formed by corroding off the debris during the generating process of tribofilm for the more serious wear under higher load; hence, wear rate is correspondingly higher. In other words, the new layer formed under higher pressure plays a role in reducing friction coefficient, which is generated by the wear debris and the effect of plastic deformation on the substrate. According to the previous results, NSMG against Si_3N_4 ball shows the better wear properties under ball-on-disk test conditions at 16-N load. To make further understanding of microstructure and formation mechanism of the friction layer, the subsurface analysis is carried out on the worn surface by crossing it perpendicularly to the sliding direction, and the location of the cross-sectional position is shown in Figure 4.12d.

Based on the aforementioned analysis and results, the microstructure of friction layer is formed during the friction and wear process. At the scale of macroscopic friction, multipoint contact friction is the prime friction formation. Consequently, low load will generate high-contact-points pressure, thereby plastic deformation that expresses as plow friction on the points [86–88]. Unlike macroscopic observation of entirely surface contact [89], as shown in Figure 4.13a, the practical contact condition at the low load is a multipoint contact model. The practical contact pairs are the bottom of ball and the asperities of the worn surface. When the low loads (2 and 6 N) are applied, the low von Mises stresses will be generated, and the formed low stresses will result in less wear material loss, which ensures low wear rate. But more asperities and flaws are retained during friction process owing to the low pressure, as shown in Figure 4.13a,

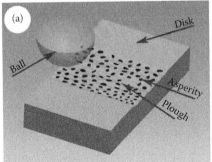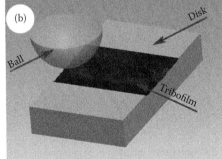

FIGURE 4.13 A schematic representation of the microstructure on the worn surface of NSMG at different loads: (a) Loads: 2 and 6 N, and (b) Loads: 12 and 16 N.

which will cause instability in friction. Consequently, the friction coefficient will keep at high level. If the high loads are applied (12 and 16 N), the von Mises stresses will create relative large deformation on the surface of contact pair; the contact points will become more or turn to surface contact. With the high von Mises stresses, asperities and flaws of contact surface will be polished and worn off; hence, wear rate increases correspondingly. However, the surface is covered by smooth and flat tribofilm, as shown in Figure 4.13b, which will make contribution to the low friction coefficient.

4.3 ALUMINUM-MATRIX COMPOSITES

4.3.1 ALUMINA MATRIX

Ceramics are relatively hard and brittle materials that exhibit superior resistance to high temperatures and severe environments compared with metals or polymers. In particular, alumina (Al_2O_3) has excellent properties, such as a high melting point, excellent wear resistance, and chemical stability. Alumina is a promising material at high temperature because of its excellent chemical stability and low price. High-performance alumina ceramic composites are potential candidates for the application of wear-resistance components because of their excellent properties [90–93]. Nevertheless, a large number of problems remain, such as high-friction coefficient of high-purity alumina ceramic. However, tribological experiments of alumina sliding against itself at high temperature show high-friction coefficient and wear rate. Solid lubrication becomes necessary to overcome this problem. To lubricate alumina ceramics, many efforts have been made in recent years. Among them, alumina matrix composite employed Ag and fluoride as solid lubricants is a successful example [16,94,95]. The Al_2O_3–Ag–CaF_2 composite exhibited a distinct improvement in wear resistance and frictional characteristics at elevated temperatures. The self-lubricating behavior was dominated by a synergistic effect. The lubricating film as a mixture of Ag and CaF_2 on friction surfaces was responsible for the reduction of friction and wear at elevated temperature.

Due to its excellent tribological properties, high thermal stability, and chemical inertness, hBN is considered to be one of the most promising solid lubricants. This versatile material is now widely used to improve the friction and wear properties of composite coatings [96,97] as well as metal and ceramic matrix self-lubricating composites. hBN is also combined with difficult-to-cut materials to enhance their machinability by dramatically decreasing the material hardness [98–101]. The low hardness, weak interfacial bonds with matrix materials, and low wettability as well as the anisotropy, agglomeration, and cleavage behaviors of hBN endow the material with good machinability.

High-performance ceramic–graphite composites are potential candidates for the application of moving components, such as cylinders, sliding bearings, and seals [15,102–104]. These materials exhibit excellent self-lubricating properties in a wide range of temperatures as the graphite can act as a lubricant to promote the formation of well-covered lubricating films on the surfaces of ceramics during sliding. The design concept of alumina/graphite laminated composites for a recent research is shown in Table 4.1. Commercially available colloidal graphite powder (≤4 μm) and

TABLE 4.1

The Chemical Component and Geometrical Parameter of Each Sample

Sample	Chemical Component		Layer-Thickness (µm)		Volume Fraction of Graphite Phase (%)	Material Type
	a	g	d_a (Spacing Among Graphite Layers)	d_g (Graphite Layer)		
A	Al$_2$O$_3$	–	–	–	0	Materials with different thickness of graphite layers
B	Al$_2$O$_3$	Graphite	440	24	5.2	
C	Al$_2$O$_3$	Graphite	440	40	8.3	
D	Al$_2$O$_3$	Graphite	440	56	11.3	
E	Al$_2$O$_3$	Graphite	440	72	14.1	
F	Al$_2$O$_3$	Graphite	440	88	16.7	
G	Al$_2$O$_3$	Graphite	440	104	19.1	
H	Al$_2$O$_3$	Graphite	440	120	21	
D	Al$_2$O$_3$	Graphite	440	56	11.3	Materials with different spacing among graphite layers
I	Al$_2$O$_3$	Graphite	1056	56	5	
J	Al$_2$O$_3$	Graphite	880	56	6	
K	Al$_2$O$_3$	Graphite	292	56	16.1	
L	Al$_2$O$_3$	Graphite	252	56	18.2	
M	Al$_2$O$_3$	Graphite	212	56	21	
H	Al$_2$O$_3$	Graphite	440	120	21	Materials with same volume fraction of graphite phase
M	Al$_2$O$_3$	Graphite	212	56	21	
N	Al$_2$O$_3$	50.0 wt.% Al$_2$O$_3$–50.0 wt.% Graphite	184	84	21	
O	–	88.3 wt.% Al$_2$O$_3$–11.7 wt.% Graphite	–	–	21	

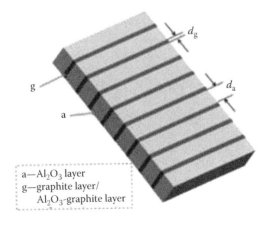

a—Al$_2$O$_3$ layer
g—graphite layer/
   Al$_2$O$_3$-graphite layer

Source: Song, J. et al., *Wear*, 338, 351–361, 2015.

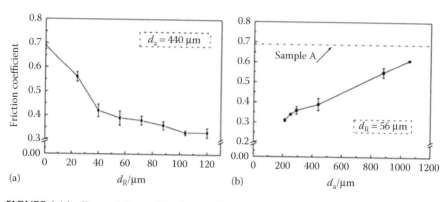

(a) $d_R/\mu m$ (b) $d_a/\mu m$

FIGURE 4.14 The variations of friction coefficient of alumina/graphite laminated composites with different thicknesses of graphite layer (a) and Al_2O_3 layer (b). (From Song, J. et al., *Wear*, 338, 351–361, 2015.)

nanosized Al_2O_3 powder (80–200 nm) modified with 5 wt.% TiO_2–CuO sintering aids (TiO_2:CuO = 4:1) were used [105].

Compared with the monolithic Al_2O_3 ceramics, alumina/graphite laminated composites exhibited excellent tribological properties. As can be seen in Figure 4.14, the friction coefficient of monolithic alumina ceramic fluctuates around 0.69, and the curve exhibits large fluctuations. However, the laminated composites have a much lower and more stable coefficient of friction than monolithic Al_2O_3 ceramics, especially the samples H and M. When the thickness of graphite layer and Al_2O_3 layer are 56 and 212 µm, respectively, the optimal friction coefficient of alumina/graphite laminated composites can be reduced to 0.31, a decrease of approximately 55% compared with that of the monolithic Al_2O_3 ceramics. In addition, the value and stability of friction coefficients of the materials are remarkably affected by the values of d_a and d_g. The friction coefficients can be reduced drastically by adjusting the values of d_a and d_g. For the samples with a constant thickness of Al_2O_3 layer (440 µm), with the increases of d_g, the friction coefficients of the laminated composites decrease rapidly at first then reduces gradually when the value of d_g exceeds 56 µm (Figure 4.14a). To reveal the effect factors on the lubricating properties of the laminated composites, the relationship of volume fraction of graphite phase and the friction coefficient of laminated composites was displayed in Figure 4.15. As shown in the figure, the volume fraction of graphite phase becomes larger with the increase of d_g (samples A–H) and decrease of d_a (samples I–M). With the increase of the volume fraction of graphite in samples, the friction coefficients of laminated composites decrease gradually. There is a relatively low value of friction coefficient (below 0.4) when the volume fraction of graphite exceeds 11%. This is mainly because the content of the graphite phase in composites directly affects the formation of the lubricating and transferring films on the sliding surfaces. On this basis, the high-friction coefficients of the samples B, J, and I (Figure 4.15) can be explained by the low content of graphite as the laminated composites are not enough for the formation of lubricating and transferring films.

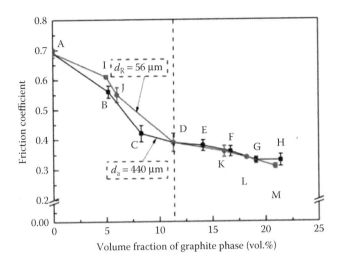

FIGURE 4.15 Relationship of the volume fraction of graphite phase and the friction coefficient of laminated composites. (From Song, J. et al., *Wear*, 338, 351–361, 2015.)

The SEM images and topographies for the worn surfaces of monolithic Al_2O_3 ceramic and alumina/graphite laminated composites are shown in Figure 4.16. Figure 4.16a–c provides the surface-state of samples A, D, and H (have different d_g) after friction test, respectively. As can be seen from Figure 4.16, the wear track and the friction coefficient of the materials are not consistent. The wear and tear of laminated samples (samples D and H) is more serious than that of monolithic Al_2O_3 ceramic, especially at the edge of the graphite layer (Figure 4.16b). The addition of a graphite lubricating phase can reduce the friction resistance of laminated composites, but the wear rate of the materials strongly depends on their structural parameters and load-bearing capacities. This can also be confirmed by the typical worn surface (Figure 4.16d–f) of samples I, K, and M (have different d_a). No matter how thick the graphite layers are, the severe wear at the edge of graphite layer is often occurred when the materials have a large thickness of Al_2O_3 layers. If the thickness of Al_2O_3 layers is extremely large, the effective lubricating films cannot be formed easily on the friction surface due to the large spacing among the graphite layers (Figure 4.16d), which will result in high friction resistance and thereby exacerbate the abrasion of materials. For the samples with a constant moderate thickness of graphite layers, the thinner thickness of Al_2O_3 layers means the smaller spacing among the graphite layers, which is conductive to the formation of lubricating and transferring films.

The SEM micrographs and EDX analysis for the worn surfaces show that the existence and distribution of graphite phase in laminated samples also have a significant impact on the wear mechanisms of friction couples. The triboinduced graphite films are beneficial to the reduction of the abrasion wear of friction couples. EDX results clearly showed that there exist graphite films on the worn surface of Al_2O_3 balls against the samples H and M, and the atomic percent of graphite on the worn surface of Al_2O_3 balls

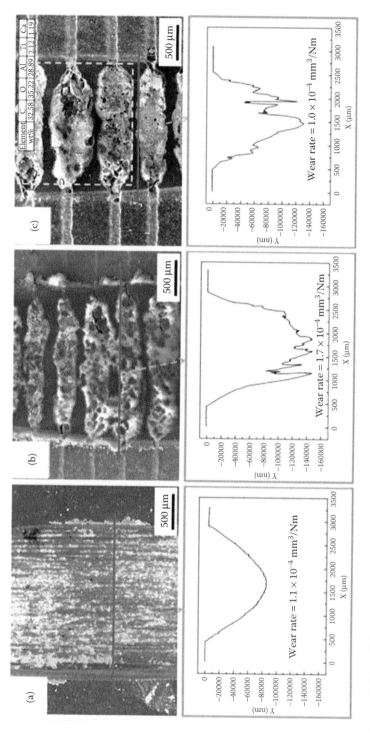

FIGURE 4.16 SEM images and topographies for worn surfaces of monolithic Al_2O_3 ceramic (a and b) and alumina/graphite laminated composites with different d_g (sample D [c and d], sample H [e and f]). (sample I [a and b], sample K [c and d]), and (sample M [e and f]). (From Song, J. et al., *Wear*, 338, 351–361, 2015.)

(*Continued*)

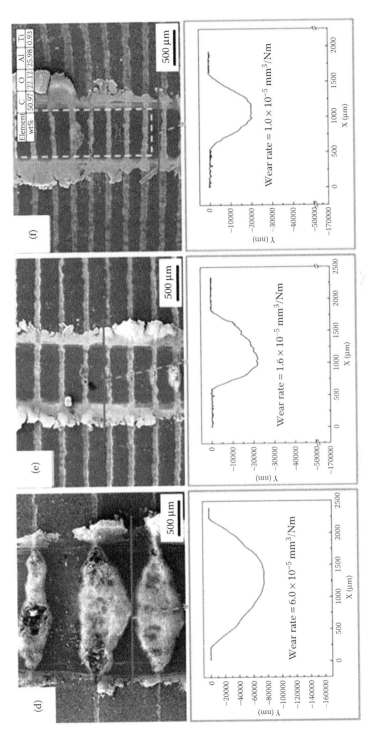

FIGURE 4.16 (Continued) SEM images and topographies for worn surfaces of monolithic Al_2O_3 ceramic (a and b) and alumina/graphite laminated composites with different d_g (sample D [c and d], sample H [e and f]). (sample I [a and b], sample K [c and d], and (sample M [e and f]). (From Song, J. et al., *Wear*, 338, 351–361, 2015.)

against the sample M can reach approximately 37%. Thus, low and stable friction coefficients can be obtained on the sliding surface. These observations supported previous results (Figure 4.14) and indicated the antifriction mechanism of laminated composites.

From the results of friction tests of alumina/graphite laminated composites, it can be concluded that the lubricating property and wear resistance of the alumina/graphite laminated composites can be drastically improved by optimizing the thickness of graphite layer and Al_2O_3 layer, and compositions of the weak layers. There is a close relationship between the friction coefficients and volume fraction of graphite phase in the composites. The laminated samples (H, M, and N) with similar content of graphite phase of approximately 21% exhibit the lowest friction coefficient of 0.31; this value is 11% lower than that of the monolithic alumina–graphite composites (sample O). However, the four materials (samples H, M, N, and O) showed a huge difference in their wear rate. The sample N exhibits the lowest wear rate of 1.5×10^{-6} mm^3/Nm. The laminated composites have a better formation style of lubricating films on the sliding surface in comparison with monolithic Al_2O_3–graphite ceramic because of the graphite phase can be easily dragged onto the friction surface from graphite layers, which will improve the concentration of lubricant on the sliding surface and reduce the friction resistance.

High-performance Al_2O_3/Mo-laminated composites are potential candidates for space applications because of their excellent self-lubricating and mechanical performance. During the past decade, inspiration from biomimetic multilayer structures like shells, which are mounts of layered ceramic composites, has been studied [106–108]. Combining the biomimetic design of ceramic materials and self-lubricating ceramic–matrix composites is a promising way to achieve the integration of mechanical and tribological properties of ceramics. According to a previous study, Al_2O_3/Mo nano-composites with laminated structure have excellent self-lubricating and mechanical properties [104]. A large number of problems remain, such as high friction coefficient of high-purity alumina ceramic and poor mechanical properties of ceramic–matrix self-lubricating materials, which limit a wider range of applications of these composites in tribological areas. According to previous studies, Al_2O_3/Mo composites with a laminated structure have excellent self-lubricating and mechanical properties [109,110]. These multilayer materials consist of a weak interfacial layer of Mo, which results in high fracture toughness and low friction coefficient.

Figure 4.17 displays the friction coefficients of Al_2O_3/Mo-laminated nanocomposites with different n at RT and 800°C, together with monolithic Al_2O_3 ($n = 0$) and pure Mo. The five laminated samples have the same d_2 of 71.5 μm. At RT, the coefficient of friction ranges from 0.77 to 0.89 and keeps at relatively high values as n changes from 0 to 24. At 800°C, the friction coefficient is between 0.86 and 0.94 when $n < 7$. With the further increase of n, the friction coefficient decreases greatly to an extremely low level and approaches 0.43 when n is 24. This friction coefficient is close to the value observed for pure Mo (0.37) under the same experimental conditions. Severe wear of Al_2O_3/Mo-laminated nanocomposites can be observed at RT, and a lot of wear debris are produced. Brittle fracture and pullout of grains can be found in the magnified picture. The total area of Mo becomes lager with the increase of n and decrease of z. The area of metal molybdenum under real contact during

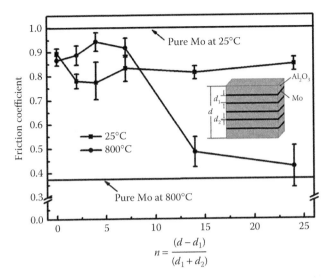

FIGURE 4.17 The effect of n on the friction coefficients of Al_2O_3/Mo-laminated nanocomposites at room temperature and 800°C. (From Fang, Y. et al., *Wear*, 320, 152–160, 2014.)

sliding is lager accordingly. Furthermore, the amount of MoO_3 generated from the oxidation reactions of metal molybdenum at 800°C increases. On the other side, d_1 decreases with the increase of n under the constant d_2. That means the distance between Mo layers is smaller. Both of the results are conductive to the formation of MoO_3 lubricating films, which take place plastic deformation during sliding and help one to decrease the friction coefficients. On this basis, the samples without self-lubricating ability at 800°C when $n < 7$ can be explained by the low content of metal molybdenum in the laminated nanocomposites, and the amount of MoO_3 generated at high temperature is not enough for the formation of lubricating films [111].

EDX of the tribofilms on the Al_2O_3 layer and the Mo layer indicates that the main composition of the tribofilms is Al_2O_3, and the composition of the tribofilms on the Al_2O_3 layers and the Mo layers are similar. EDX of the area uncovered with tribofilms correspond to the Mo layer and the Al_2O_3 layer, respectively. The small amount of Ti element presented at the area uncovered with tribofilm on the Al_2O_3 layer is due to that TiO_2 was added in Al_2O_3 as sintering aids. There is no Ti element detected in the tribofilms that shows the wear debris is mainly generated from upper samples (Al_2O_3 pins).

To overcome the anisotropic of Al_2O_3/Mo-laminated composites, alumina/molybdenum fibrous monolithic ceramics were prepared in compliance with the design principle of bionic bamboo [112]. Therefore, improving high-toughness ceramic–matrix self-lubricating materials for actual applications is significant. Combining the bionic design of ceramic materials and self-lubricating ceramic–matrix composites with excellent lubricating property is a promising way to integrate mechanical and tribological properties [30,104,111,113]. Figure 4.18 displays a summary of the average friction coefficients of alumina/molybdenum fibrous monoliths ceramic in

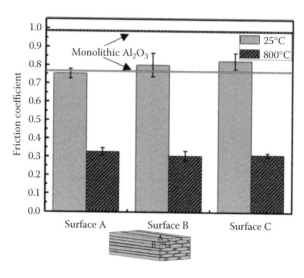

FIGURE 4.18 Average value of friction coefficients of alumina/molybdenum fibrous monoliths ceramic and monolithic Al_2O_3 at 25°C and 800°C. (From Su, Y. et al., *Tribol. Lett.*, 61, 9, 2016.)

different surfaces (A, B, and C) at RT and 800°C, together with monolithic Al_2O_3. At RT, the friction coefficient ranges from 0.75 to 0.83 and stays at relatively high values on any surface. At 800°C, however, the friction coefficient decreases greatly to an extremely low level and approaches 0.30 on every surface, which is much lower (approximately 0.31 times) than that of the monolithic composites. This value is also 0.60–0.90 times lower than that of the Al_2O_3/Mo-laminated composites with different structural parameters [111]. Severe wear and rough morphologies with lots of debris can be observed at RT, which indicates that the brittle microfracture is the major wear mechanism. Compared with the surfaces at RT, the wear scar at 800°C is much smoother without distinct wear debris and presents the property of plastic deformation forming successive lubricating films on it. The XRD patterns of the worn surface after the friction test at RT are covered with α-Al_2O_3 and metal Mo, whereas the worn surface at 800°C is covered with α-Al_2O_3, MoO_3, $MoO_{2.8}$ and a small amount of metal Mo, indicating that the oxidation of molybdenum and molybdenum oxides during sliding is an important factor in reducing friction coefficients, which is similar to our previous results.

Alumina exhibits lower fracture toughness than other ceramics. Recently, researchers have reported that fracture toughness can be improved by the addition of second phase particles, such as platelets, whiskers, and fibers [114–116]. The fracture toughness and flexural strength can also be enhanced by dispersing nanometer-sized secondary phase materials [117,118]. The use of tetragonal ZrO_2 (t-ZrO_2) has been shown to improve the mechanical properties of Al_2O_3 ceramics, thereby producing a ceramic known as zirconia toughened alumina (ZTA) [119]. The toughening mechanism of ZTA is based on the stress-induced martensitic transformation and microcrack toughening. The fracture toughness and flexural strength of ZTA are 7 MPa m$^{1/2}$

and 910 MPa, respectively [120]. The tribological properties of self-lubricating Al_2O_3–ZrO_2 composites containing such fluorides as CaF_2 and BaF_2 were investigated. The composites were fabricated using a pulse electric current sintering (PECS) technique. Aluminum oxide (α-Al_2O_3, AKP53, Sumitomo Chemical Ltd. Co., Japan), 3 mol% yttrium oxide stabilized zirconium oxide (ZrO_2, TZ-3Y-E, Tosoh Corporation, Japan), calcium fluoride (CaF_2, High Purity Chemicals, Japan), and barium fluoride (BaF_2, High Purity Chemicals, Japan) were used as the starting materials. The ZrO_2 and fluoride contents were set to 15 wt.% and 3 wt.%, respectively [121]. The coefficient of friction and wear rate of the Al_2O_3–ZrO_2 composites are shown in Figures 4.19 and 4.20. The coefficient of friction of Al_2O_3–ZrO_2 composite containing solid lubricants was lower than that of the matrix Al_2O_3–ZrO_2 composite. The coefficient of friction of the latter material decreased from 0.5 to 0.35 when the sliding time was increased. At the same time, the friction coefficient of the CaF_2 containing Al_2O_3–ZrO_2 composites

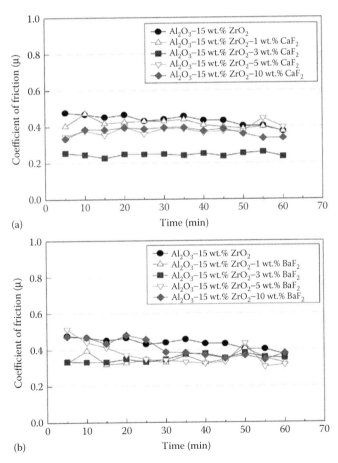

FIGURE 4.19 The coefficient of friction of Al_2O_3–15 wt.% ZrO_2 composites containing different solid lubricants: (a) CaF_2 and (b) BaF_2. (From Kim, S.-H. and Lee, S.W., *Ceram. Inter.*, 40, 779–790, 2014.)

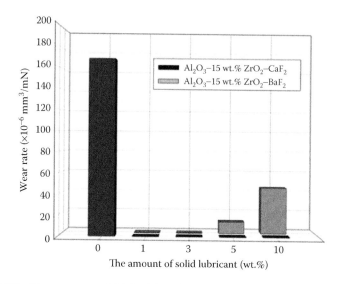

FIGURE 4.20 The wear rates of Al_2O_3–15 wt.% ZrO_2 composites containing different solid lubricant. (From Kim, S.-H. and Lee, S.W., *Ceram. Inter.*, 40, 779–790, 2014.)

remained constant in the range of 0.3–0.5, except for the specimen containing 3 wt.% CaF_2, which was approximately 0.25. In contrast, the coefficient of friction of the BaF_2 containing Al_2O_3–ZrO_2 composite varied as a function of sliding time. As the sliding time increased, the coefficient of friction exhibited two different regimes with varying BaF_2 content. The coefficient of friction of composites containing a low fraction of BaF_2 (1 wt.% and 3 wt.% BaF_2) increased slightly, whereas those containing a high fraction of BaF_2 (5 wt.% and 10 wt.% BaF_2) decreased to approximately 0.1. The friction coefficient of composites containing solid lubricants was similar to that of the matrix composite as a function of sliding time. Figure 4.20 shows the wear rates of the Al_2O_3–ZrO_2 composites as a function of solid lubricants content. The wear rate of the solid lubricant containing Al_2O_3–ZrO_2 composites was lower than that of the matrix Al_2O_3–ZrO_2 composite. As the amount of CaF_2 increased, the wear rate of the Al_2O_3–ZrO_2 composite remained constant between 0.25 and 0.40×10^{-6} mm³/mN, which was several times higher than that of other composites. In contrast, the wear rate of the BaF_2-containing Al_2O_3–ZrO_2 composite increased with increasing BaF_2 content from 0.36 to 45.05×10^{-6} mm³/mN. In particular, the wear rate of the composites increased dramatically above 5 wt.% BaF_2. Furthermore, the wear rate of the BaF_2-containing Al_2O_3–ZrO_2 composite was closely related to the initial friction coefficient. The initial friction coefficient of the low fraction composites (1 wt.% and 3 wt.% BaF_2) was approximately 0.33–0.34, whereas that of the high fraction composites (5 wt.% and 10 wt.% BaF_2) were approximately 0.47–0.51. Conversely, the steady-state friction coefficient of the low and high fraction composites was similar in the range of 0.33–0.38. In the case of BaF_2-containing composites, the wear rates were closely related to the friction coefficient behavior of the material. The initial friction coefficient was affected by the presence of fracture or microcracks. The high wear rate of the Al_2O_3–ZrO_2 composite containing 10 wt.% BaF_2 was caused by the long dwell times of the high initial friction coefficient. In contrast, the wear rate of the Al_2O_3–ZrO_2 composite containing 5 wt.% BaF_2 was lower than that of composite containing 10 wt.% BaF_2. The friction coefficient of the Al_2O_3–ZrO_2 composite containing 5 wt.% BaF_2 decreased gradually with increasing sliding time.

Fine wear debris can be observed in the worn surface of the composite, as well as the wear debris-filled pores. The smooth-worn surface of the CaF_2-containing Al_2O_3–ZrO_2 composite led to a decrease in the friction coefficient and wear rate. In contrast, the surface roughness of the BaF_2-containing Al_2O_3–ZrO_2 composite increased with increasing BaF_2 content. As the amount of BaF_2 increased, the number of microcracks on the worn surface also increased. The worn surface of the low weight fraction composites (1 wt.% and 3 wt.% BaF_2) exhibited similar trends to those of the CaF_2-containing composites, in which fine wear debris were observed on the worn surface. The low wear rates and friction coefficients of the composites containing a low weight fraction of BaF_2 were due to the smooth-worn surface.

As Al_2O_3/TiC composites [122,123] have high hardness and intensity values and Al_2O_3/TiC/CaF_2 composites [124] possess better tribological properties, Al_2O_3/TiC–Al_2O_3/TiC/CaF_2 self-lubricating laminated ceramic materials were proposed and the mechanical properties, friction, and wear performance under dry friction conditions were investigated. The main powders used to fabricate the laminated ceramic composites were Al_2O_3, TiC, CaF_2, Mo, and Ni powder. The density of high-purity α-Al_2O_3 powder with an average particle size of 0.5 μm was 3.99 g/cm³ and the density of high-purity TiC powder with an average particle size of 0.5 μm was 4.25 g/cm³. The purity of CaF_2 powder was more than 98.5%, and its density was 3.18 g/cm³. The purity of Mo powder and Ni powder was more than 99% [124].

Figure 4.21a and b illustrates the effect of loads on friction coefficient and wear rates of Al_2O_3/TiC and Al_2O_3/TiC/CaF_2 self-lubricating laminated ceramic composites, respectively. Obviously, the friction coefficient and wear rates show downward trend with an increase in the load under certain rotation speed. When the rotation speed was 100 r/min and the load was in a lower scales (50–150 N), both of the values of friction coefficient and wear rates were in a smaller location. When the load was 250 N, friction coefficient decreased to 0.46 and the corresponding wear rate was 1.03×10^{-8} cm³/Nm. Under the same speed conditions with a smaller load, the contact surfaces with tiny and hard particles or micro convex peaks had an adverse effect to friction process. These tiny particles acted as cutting tools for quenching 45 # steel materials, resulting in mass loss of friction pairs. The mechanism shows the reason why the friction pair exerts high friction coefficient and wear rates in a smaller load. With the increase of loads, temperature of friction surface increases and plastic deformation is generated at the contact areas on friction surface, which improve friction conditions contributing to the decrease of friction coefficient and wear rates. Furthermore, improvement of lubrication effect of Al_2O_3/TiC/CaF_2 layers caused by the rising temperature also has a positive function. Figure 4.21c and d illustrates the effect of the rotation speeds on friction coefficient and wear rates of Al_2O_3/TiC and Al_2O_3/TiC/CaF_2, self-lubricating laminated ceramic composites when the load was 200 N, respectively. Obviously, the two figures show that the

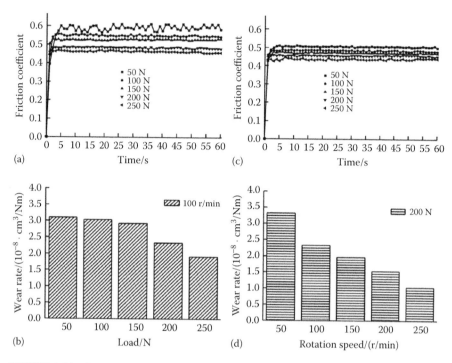

FIGURE 4.21 The effect of loads on (a) COF and (b) wear rates of $Al_2O_3/TiC-Al_2O_3/TiC/CaF_2$ and the effect of sliding velocity on (c) COF and (d) wear rates of $Al_2O_3/TiC-Al_2O_3/TiC/CaF_2$.

friction coefficient, and wear rates decrease with the rise of rotation speeds. When the rotation speed is 50 r/min, the friction coefficient and wear rates are 0.51 and 3.78×10^{-8} cm³/Nm. Besides, when the rotation speed was 50 r/min, the corresponding values reduced to 0.44 and 1.91×10^{-8} cm³/Nm. Under the same load condition with a lower speed, the temperature of friction surface was not high enough to make CaF_2 particles change from brittle state to the plastic state. CaF_2 particles are difficult to separate out from matrix and adhere to the friction surface, so the complete lubrication films could not form. When speed increases, the contact areas are under high temperature and pressure conditions that are necessary for CaF_2 particles to transform into plastic state. Meantime, CaF_2 particles are squeezed out from matrix materials because of high coefficient of thermal expansion. Due to dragging effect in the process of friction, solid lubricant CaF_2 particles are towed and covered on the friction surface, improving the friction and wear conditions. Because of the tribofilms formed by CaF_2 particles, the friction coefficient and wear rates at high rotation speed are smaller than those at low rotation speed.

4.3.2 Aluminum Nitride Matrix

Ceramic cutting tools can be used for both rough cutting and finishing operation. Despite progress in nonconventional manufacturing technologies, metal cutting still plays a major role in the production of machines, products, apparatus, and goods for everyday use. It makes possible achievement of the highest geometric accuracy of products and produces best surface finish. Hard turning using ceramic insert has advantages of increased productivity, reduced setup times, and surface finish closer to grinding. Xiang et al. [125] report that AlN (64%), TiB_2 (30%), and additive Y_2O_3 (6%) fabricated by SPS at 1560°C with 30 MPa uniaxial load for 4 min have hardness and toughness nearly 13.5 GPa and 4.8 MPa.m$^{1/2}$.

4.4   TITANIUM MATRIX

Thanks to their remarkable properties such as low density, high specific strength, elastic modulus retention, high melting point, good oxidation resistance at elevated temperature high dimensional, and good environmental stabilities, TiAl-based inter-metallics have been widely selected as ideal high-temperature structural and engine materials [126–130]. In recent years, TiAl turbocharger turbine wheel has been used for commercial cars [126], and other TiAl products, such as low-pressure turbine blade, corner-beam, transition-duct beam, and so on, have also reached the engineering technology level [129]. However, their applications are significantly hindered because of the poor ductility at ambient temperature and low creep resistance at elevated temperature [131]. Because sliding contact occurs in many potential applications of TiAl-based intermetallics, which is associated with friction and wear, it is vital to study their tribological behavior under sliding conditions. TiAl alloys are regarded as potential high-temperature self-lubrication alloy matrix [132,133].

Li et al. [134] studied the sliding wear of TiAl intermetallics against steel and ceramics of Al_2O_3, Si_3N_4, and WC/Co. They found that sliding wear property of titanium aluminide was strongly dependent on the counterface materials. Cheng et al. [135] investigated the effect of TiB_2 on dry-sliding tribological properties of TiAl intermetallics. They observed that the wear resistance increased dramatically, but the friction coefficient was not related to the addition of TiB_2. In addition, several investigations showed that titanium aluminides had poor tribological behaviors that would limit their applications [136,137]. Hence, the addition of effective solid lubricants becomes an excellent solution to obtain the TiAl matrix materials with good tribological behaviors over the wide temperature range. Cheng et al. [138] investigated the tribological behavior of a Ti–46Al–2Cr–2Nb alloy under liquid paraffin lubrication against AISI 52100 steel ball in ambient environment and at varying loads and sliding speeds. Nevertheless, only several studies have been done on the tribological behavior of TiAl intermetallic materials at elevated temperatures. The fretting wear behavior of a Ti–48Al–2Cr–2Nb alloy was investigated in air from RT to 600°C [139]. Tribological behavior of a Ti–46Al–2Cr–2Nb alloy was investigated against a Si_3N_4 ceramic ball at a constant speed of 0.188 m/s and an applied load of 10 N from 20°C to 900°C [138,140]. Consequently, it is necessary to further investigate the elevated temperature tribological behavior of TiAl-based intermetallic materials.

It is well known that the layered carbide-like ternary compound Ti_3SiC_2 has ignited a worldwide attention due to its remarkable properties. It displays not only metal properties, including fine thermal conductance, better conductivity, high elastic, and shear modulus, easy machining property, and plasticity at high temperature, but also ceramic properties, that is, high yield strength, high melting point, and high thermal stability [141–145]. A number of studies on its tribological behavior have also been reported, which indicate that Ti_3SiC_2 has good tribological properties [146–148]. Hence, Ti_3SiC_2 may be an ideal candidate as a structural ceramic for high temperature and a new solid lubricant material at high temperatures. To the authors' knowledge, reports on Ti_3SiC_2 as a solid high-temperature lubricant in composites are very rare, especially in TiAl matrix composite. Thus, it is valuable to explore the high-temperature lubrication mechanism of Ti_3SiC_2 in TiAl-matrix composite. Ti_3SiC_2 combines many of the best properties of both ceramics and metals [49]. Like graphite and MoS_2, Ti_3SiC_2 has a hexagonal structure with a space group of P63/mmc. Besides, it is readily machined as graphite and possesses excellent thermostability. Consequently, Ti_3SiC_2 can be a new solid self-lubricating material at high temperatures [47]. The good tribological properties are attributed to the formation of a lubricious oxide film on the Ti_3SiC_2 [149]. Gupta et al. [150] found that when tested in air at 550°C, the friction coefficient for the Ti_3SiC_2 sliding against Al_2O_3 was the lowest one of 0.4. The superiority of Ti_3SiC_2 is not just its lubrication, and TiC particles that the impurity distributes in Ti_3SiC_2 can reinforce materials are expected to process excellently both abrasive and adhesive wear resistance because of high hardness of TiC reinforcement [63].

A research studied the friction and wear properties of Ti_3SiC_2/TiAl composite (TTC) produced by *in situ* technique using SPS against Si_3N_4 ceramic ball from RT to 800°C [151]. Figure 4.22a and b exhibits the friction coefficient and wear rate of TiAl-based alloy (TA) and TTC as a function of temperature. It is clear that the variation trends of both friction coefficient and wear rate with temperature for TA and TTC are the same. The friction coefficient and wear rate first increase and reach to the highest point at 400°C, then decrease to the lowest value with increasing test temperature up to 800°C. Furthermore, it can be observed that the friction coefficients and wear rates of TTC are comparable to those of TA at 25°C–400°C, whereas the gaps in friction coefficients and wear rates between TTC and TA beyond 400°C significantly appear. The reason for these is that Ti_3SiC_2 being a newly known high-temperature lubricant plays a dominant role in friction-reduction and antiwear properties at high temperatures of 600°C and 800°C. As shown in Figure 4.22a, the friction coefficients of TA and TTC at the test temperatures from 25°C to 400°C are in the range of 0.44–0.52. The friction coefficients of TTC are observed to remarkably decrease from 0.46 to 0.34 at 800°C after adding the Ti_3SiC_2 lubricant. Meanwhile, as is seen obviously from Figure 4.22b, the wear rates for TA and TTC at the same test temperatures from 25°C to 400°C are comparable. For the high temperatures of 600°C and 800°C, the wear rates of TTC fall from 3.42 to 1.21×10^{-4} mm^3/Nm and from 2.65 to 0.85×10^{-4} mm^3/Nm by the addition of Ti_3SiC_2 lubricant, respectively. From the above analyses, it can be concluded that TTC when compared with TA only exhibits excellent tribological properties at the elevated temperatures of 600°C and 800°C. It indicates that Ti_3SiC_2 can actually act as an ideal high-temperature solid lubricant in TTC but does not well work at 25°C–400°C. We will emphatically analyze and discuss the lubrication mechanism of Ti_3SiC_2, which governs the tribological behaviors of TTC at elevated temperatures.

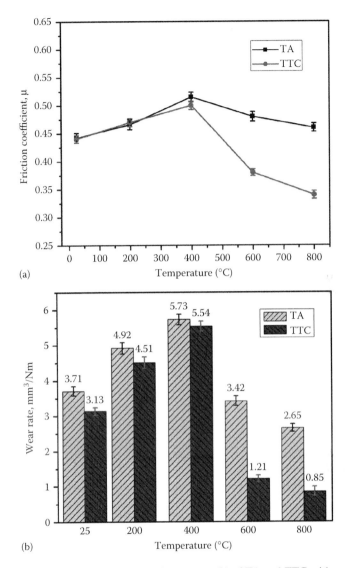

FIGURE 4.22 Friction coefficient (a) and wear rate (b) of TA and TTC with temperature. (From Xu, Z. et al., *J. Mater. Engi. Perform.*, 23, 2255–2264, 2014.)

It is clear to see from electron probe micro-analyzer (EPMA) morphologies of worn surfaces of TA and TTC that the morphology of worn surface of TTC is different from that of TA. Some wider but shallow grooves characterizing the abrasive nature of wear can be observed on the worn surface of TA that is covered with local compacted tribofilms. The elemental compositions on the worn surface ($Ra = 1.943$ μm) of TAs indicate that the compacted tribofilms mainly consist of Al–Ti oxides, whereas it shows a smooth surface ($Ra = 0.733$ μm) covered with compacted tribofilms, and no such grooves can be seen on the worn surface of TTC. Moreover, the extent of covered tribofilms in TTC is larger in comparison with that observed in TA. Besides, the smooth surface is covered with loose wear debris at a small location that appears bright. The EDX analysis of TTC indicates that the compacted tribofilm is an oxide-rich tribofilm mainly consisting of Al–Ti–Si–oxides, whereas the wear debris mainly consists of Al–Ti–Si–oxides. It is well known that it was quite difficult to directly obtain an XRD pattern of the tribofilm on the worn surface, because the X-ray can readily penetrate the tribofilm and reach into the surface of the matrix. Hence, to further identify the information of phase composition of the tribofilm on the worn surface, the XPS analysis is used herein. From the XPS results, it can be found that the peaks assigned to TiO_2, SiO_2, and Al_2O_3 are detected. The presence of Si–oxide implies that Ti_3SiC_2 has been decomposed and oxidized during the friction process. It is clear from the XPS results that the tribofilms formed on the worn surface of TTC mainly consist of Al–Ti–Si–oxides, which reduces TTC–Si_3N_4 contact and provides the low shear strength junctions at the interface, thus reducing the plowing component of friction force which in turn reduces the wear rate and friction coefficient. This should be responsible for the lower friction coefficient and less wear rate of TTC at 600°C, as seen in Figure 4.22a and b.

BaF_2/CaF_2 eutectic (BaF_2 38 wt.% CaF_2), which has wide working temperature range and can lubricate effectively above 400°C, has been widely used to lubricate many wear-resistant matrices [152,153]. The fluoride eutectic undergoes a brittle to ductile transition at about 400°C, resulting in a reduction in shear strength and increasing its effectiveness as a lubricant [45]. TiAl matrix self-lubricating composites (TMSC) containing varying amounts of Ag, Ti_3SiC_2, and BaF_2/CaF_2 eutectic (ATBC), whose weight ratio is 1:1:1, are prepared by *in situ* technique using SPS to reach the lubrication over a wide temperature range and expand the application range of TiAl intermetallic matrix composites [154]. Table 4.2 gives the compositions of the as-prepared specimens. Figure 4.23 shows the variations of friction coefficients of TMSC added with different ATBC contents over a wide temperature range from RT to 600°C at the condition of 10 N–0.234 m/s. It can be found that the friction coefficients of TA, TB, TC, and TD almost have the similar variation tend.

TABLE 4.2
Compositions, Vicker's Microhardness and Density of the As-Prepared TiAl-Matrix Self-Lubricating Composites (TMSC)

Samples	Compositions (wt.%)	Hardness (HV1)	Density (g/mm³)	Relative Density (%)
TA	TiAl	557 ± 33	3.85 ± 0.02	98.7
TB	TiAl–3Ag–3Ti$_3$SiC$_2$–3BaF$_2$/ CaF$_2$ eutectic	609 ± 26	3.95 ± 0.02	99.2
TC	TiAl–5Ag–5Ti$_3$SiC$_2$–5BaF$_2$/ CaF$_2$ eutectic	595 ± 23	4.00 ± 0.02	99.5
TD	TiAl–7Ag–7Ti$_3$SiC$_2$–7BaF$_2$/ CaF$_2$ eutectic	531 ± 45	3.98 ± 0.02	98

Source: Shi, X. et al., *Mater. Des.*, 53, 620–633, 2014.

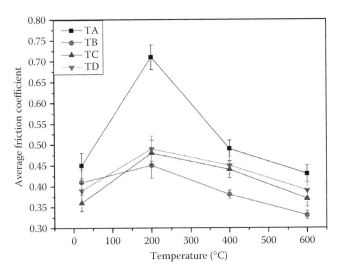

FIGURE 4.23 Variations of friction coefficients of TMSC added with different ATBC contents with temperatures. (From Shi, X. et al., *Mater. Des.*, 53, 620–633, 2014.)

The friction coefficients initially increase from RT to 200°C and then gradually decrease with the increase in temperature. TA, TB, TC, and TD obtain the maximum friction coefficient values at 200°C, which are about 0.71, 0.45, 0.48, and 0.49, respectively. The friction coefficients of TA, TB, TC, and TD reach a minimum value at 600°C, which are about 0.43, 0.33, 0.37, and 0.39, respectively. The friction coefficient of TA from RT to 600°C is the highest of the four samples and reaches its maximum value at 200°C, while reaches its minimum value at 600°C. TC obtains the lowest friction coefficient of about 0.36 among the four samples at RT, whereas the friction coefficient of TB is the lowest among the four samples at 200°C, 400°C, and 600°C, which is 0.45, 0.38, and 0.33, respectively. Figure 4.24 shows the variations of wear rates of TMSC added with different ATBC contents with temperatures after tests. It can be seen that the wear rates of TMSC added with different ATBC contents show the different variation trends with the increase in temperature. The wear rate of TA initially increases at 200°C, and then gradually decreases with the increase in temperature, until reaches the lowest value of 4.4×10^{-4} mm³/Nm at 600°C. The wear rate of TB increases with the increase in temperature, until reaches the largest value of 3.64×10^{-4} mm³/Nm at 400°C, and then decreases to 3.26×10^{-4} mm³/Nm at 600°C. The wear rate of TC initially decreases at 200°C, and then gradually increases with the increase in temperature, until reaches the lowest value of 3.82×10^{-4} mm³/Nm at 600°C. The wear rate of TD slips to 4.0×10^{-4} mm³/Nm at 600°C from 5.21×10^{-4} mm³/Nm at RT.

For TA, at RT, abundant wear particles and some pits are found on the worn surface as shown in Figure 4.25a, it reveals that the wear mechanism is the typical abrasive wear. At 200°C, as shown in Figure 4.25b, the coarse and deep parallel grooves appear. In addition, a few delamination pits are also found. It reveals that the wear mechanisms are mainly microcutting wearing and furrow. At 400°C, not only

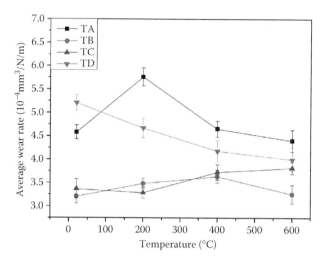

FIGURE 4.24 Variations of wear rates of TMSC added with different ATBC contents with temperatures. (From Shi, X. et al., *Mater. Des.*, 53, 620–633, 2014.)

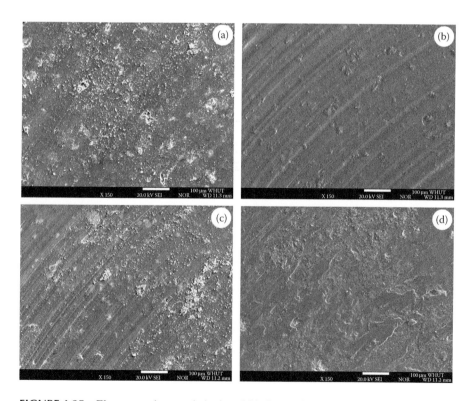

FIGURE 4.25 Electron probe morphologies of friction surfaces of TA over a wide temperature range from RT to 600°C after tests: (a) RT, (b) 200°C, (c) 400°C, and (d) 600°C. (From Shi, X. et al., *Mater. Des.*, 53, 620–633, 2014.)

abundant wear particles and some pits, but also coarse and deep parallel grooves are found on the worn surface as observed in Figure 4.25c. As temperature increases to 600°C, the worn surface of TA is covered by a tribofilm that is deeply torn as can be seen from Figure 4.25d and the serious delamination can be also observed on the deformed surface. The results showed that as temperature reached 400°C and 600°C, the worn surfaces were composed of a large amount of TiO_2 and Al_2O_3, which had a contrary effect to the friction coefficient. Furthermore, it was found that oxide might be formed by oxidation of the metal asperities while in contact, and the extent of such oxidation in one traversal was dependent on the temperatures developed at the asperity contacts. For TB, it can be seen that the morphologies of the wear surfaces at testing temperatures from RT to 400°C are almost similar. At RT, the fine and shallow parallel grooves, and some peeling pits are found on the worn surface shown in Figure 4.26a. With the increase in test temperature, the worn surfaces become much smoother, but the coarser and deeper parallel grooves appear when compared with that of RT as shown in Figure 4.26b and c. A few peeling pits are also found. In addition, it seems that there are the thin tribofilms existing on the surface of TB as the test temperature reaches 400°C. It means that the main wear mechanisms are furrow and delamination at temperatures from RT to 400°C. As temperature reaches 600°C, the

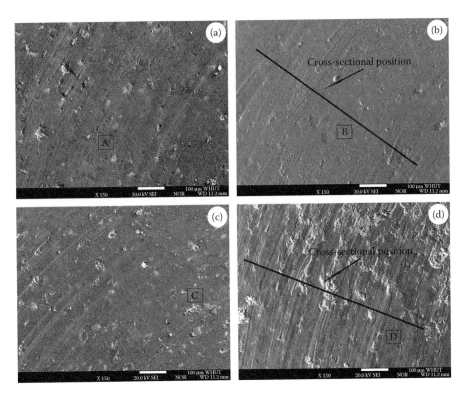

FIGURE 4.26 Electron probe morphologies of friction surfaces of TB over a wide temperature range from RT to 600°C: (a) RT, (b) 200°C, (c) 400°C, and (d) 600°C. (From Shi, X. et al., *Mater. Des.*, 53, 620–633, 2014.)

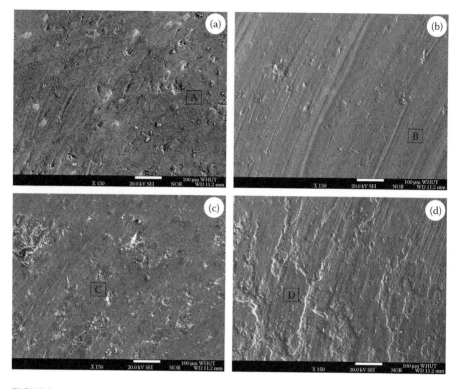

FIGURE 4.27 Electron probe morphologies of friction surfaces of TC over a wide temperature range from RT to 600°C: (a) RT, (b) 200°C, (c) 400°C, and (d) 600°C. (From Shi, X. et al., *Mater. Des.*, 53, 620–633, 2014.)

tribofilm of TB as shown in Figure 4.26d becomes much denser when compared with that of TA as shown in Figure 4.25d and the serious delamination can also be observed on the deformed surface. For TC, it can be seen that the morphologies of the wear surfaces at testing temperatures from RT to 600°C are almost similar to that of TB. There are grooves, wear debris, and peeling pits on the worn surfaces at temperatures from RT to 400°C as shown in Figure 4.27a–c. It means that the main wear mechanisms are furrow and delamination at temperatures from RT to 400°C. As temperature increases to 600°C, the wear track shows the formation of discontinuous smooth island-like (patchy) films in addition to a rougher surface. The wear morphologies of the wear surface as shown in Figure 4.27d undoubtedly suggest that the wear mechanism is the dominant surface deformation, to know the effect of lubrication phases on the friction and wear behavior. The presence of Ag element should be mainly responsible for the lower friction coefficient from RT to 400°C. Results of EDX showed that there are C, O, F, Al, Si, Ca, Ti, and Ba elements existing. As depicted in Figure 4.23, there is a decrease in friction coefficient of TC even after 400°C when the lubricating effect of Ag starts diminishing. It may be attributed to the antifriction effect of BaF_2/CaF_2 eutectic and Ti_3SiC_2, which have wide working temperature range and can lubricate effectively above 400°C. For TD, at temperatures from RT to 200°C, there are

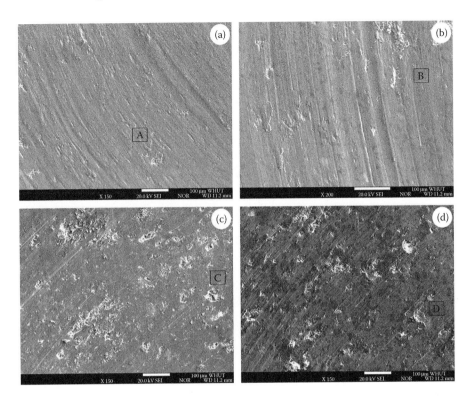

FIGURE 4.28 Electron probe morphologies of friction surfaces of TD over a wide temperature range from RT to 600°C: (a) RT, (b) 200°C, (c) 400°C, and (d) 600°C. (From Shi, X. et al., *Mater. Des.*, 53, 620–633, 2014.)

deep parallel grooves and peeling pits on the worn surfaces as shown in Figure 4.28a and b. The main wear mechanisms are furrow and delamination. At temperatures from 400°C to 600°C, the worn surfaces shown in Figure 4.28c and d become much smoother when compared with those of RT and 200°C. In addition, it seems that there are thin films existing on the worn surfaces. Furthermore, there are a considerable amount of peeling pits, which imply that the main wear mechanism is a synthetically combined function of microcutting wearing and surface peeling. They show that the compositions mainly are the elements of C, O, F, Al, Si, Ti, Ag, Cr, and Nb. In addition, at 400°C, Ba and Ca elements can also be found. It reveals that the lubricants of Ag, Ti_3SiC_2, and BaF_2/CaF_2 eutectic have played a role at the sliding interface in the process of sliding wear test over the temperature range from RT to 600°C.

4.5 SILICON NITRIDE MATRIX

Si_3N_4-based ceramics are potential substitutes for more traditional materials for these specific applications due to their high hardness, excellent chemical and mechanical stability under a broad range of temperatures, low density, low thermal expansion, and high specific stiffness [155]. The incorporation of solid lubricants is a goal to further enhance the tribological performance of Si_3N_4 [20,156–160]. The published papers indicated that cesium compounds are exceptional promises as high-temperature lubricants for Si_3N_4 ceramic. Cesium compound provided favorable lubrication on Si_3N_4 from RT to 750°C, especially with an average value of 0.03 at 600°C. The synergistic chemical reactions occurred between the cesium compounds, Na_2SiO_3, and the Si_3N_4 surface to provide the remarkable performance. In addition, just like most of the ceramic materials, the machinability of Si_3N_4 ceramics is extremely poor [161]. It is well known that hBN offers a number of interesting properties such as lubrication action, low hardness, and low friction coefficient [162]. hBN has also excellent machinability due to a plate-like structure similar to that of graphite. To improve the machinability, fracture toughness and tribological properties (by the selective oxidation of the compounds) of Si_3N_4 ceramics at room and elevated temperatures, hBN particles were introduced as second-phase dispersions into the Si_3N_4 matrix by many researchers [20,163–166]. In Si_3N_4/BN composite ceramics, the cleavage behavior of plate-like structured BN particles endowed the material with good machinability together with superior thermal shock resistance [167]. Silicon nitride materials containing 1–5 wt.% of hBN (microsized or nanosized) were prepared by hot-isostatic pressing at 1700°C for 3 h. Effect of hBN content on microstructure, mechanical, and tribological properties has been investigated. Both the silicon nitride and hBN showed relatively low friction and wear. hBN as a solid lubricant has high performance only at high relative humidity [168]. The formation of oxide or hydrated layers (H_3BO_3 and $BN(H_2O)_x$) is reported to have a beneficial effect on the tribological performance of Si_3N_4–BN composites at RT, reducing the wear coefficient by one order of magnitude to $k = 10^{-6}$ mm³/Nm, relative to the matrix material [156]. High tribological performance of hBN is reported to be controlled by a basal plane slip or tribological products such as B_2O_3 [169]. To find out the effect of size of solid lubricant, two types of materials were prepared. First group was samples when microsized Si_3N_4 particles were mixed with microsized hBN particles (Si_3N_4/BN micro/microcomposites). Second type was samples when microsized Si_3N_4 particles were mixed together with nanosized hBN particles (Si_3N_4/BN micro/nanocomposites) [99].

Figure 4.29 summarizes the average values of the coefficient of friction for all experimental composites in contact with Si_3N_4 ball. COF of composites are similar to the reference Si_3N_4. The COF of monolithic Si_3N_4 material was around 0.7, which is similar to other investigation. It can be seen that for used amounts of boron nitride (1, 3, and 5 wt.%), the COF remained basically the same. The COF of Si_3N_4/BN micro/nanocomposites varied from 0.64 to 0.73, and COF of Si_3N_4/BN micro/microcomposites were between 0.69 and 0.74. hBN has a lower friction coefficient and lubrication action; thereby the COF of Si_3N_4/hBN ceramic composites should decrease with increasing BN content, but in our study, no significant lubrication effect was observed. Investigation of Wei et al. [162] confirmed lower COF of Si_3N_4/BN in the case of BN addition higher than 10 wt.%. Chen et al. [170] showed that the addition of hBN to Si_3N_4 resulted in decrease of COF from 0.95 for Si_3N_4 against stainless steel to 0.03 for Si_3N_4-30% hBN against stainless steel. Skopp and Woydt [163] investigated the tribological performance of Si_3N_4/BN composites under unlubricated sliding conditions. Their results revealed that COF of Si_3N_4 is between 0.4 and 0.9. When the addition amount of hBN raised to 20 wt.%, the friction coefficient decreased to a range between 0.1 and 0.3 at RT. Carrapichano et al. [20] also

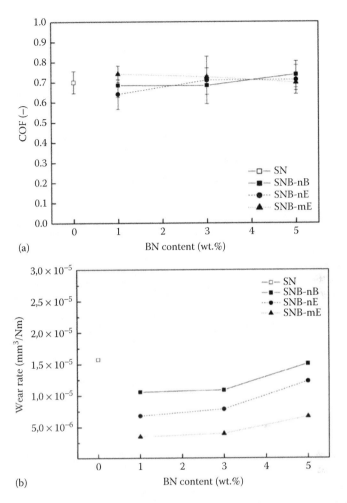

(a)

(b)

FIGURE 4.29 (a,b) Effect of hBN content on the coefficient of friction of Si_3N_4/BN composites (SNB-nB: NanoBN addition at the beginning of milling, SNB-nE: NanoBN addition at the end of milling, SNB-mE: MicroBN addition at the end of milling). (From Kovalčíková, A. et al., *J. Euro. Ceram. Soc.*, 34, 3319–3328, 2014.)

reported a slight reduction of the friction coefficient from 0.82 for Si_3N_4 to 0.67 for Si_3N_4-10 vol% hBN. Another report suggests that the COF is not decreased following the addition of hBN into Si_3N_4 at less than 30% [171]. For all experimental materials, the wear rates were lower than for the monolithic Si_3N_4 (Figure 4.29). The Si_3N_4/BN micro/microcomposites had the values of specific wear rate one order of magnitude lower (3.5×10^{-6}–6.7×10^{-6} mm^3/Nm) compared with monolithic Si_3N_4 (1.6×10^{-5} mm^3/Nm). The tendency suggests that larger BN particles coincide with better wear resistance. The lowest wear resistance had material with 5 wt.% nano BN particles prepared by first approach with the wear rate close to that of monolithic Si_3N_4. From Figure 4.29, it is clearly visible that the positive effect of BN on friction is accompanied by an increase in wear resistance only when low BN content (up to 5 wt.%) is added to the Si_3N_4. At 5 wt.% of BN particles, the specific wear rates slowly increased. Our results are similar with other authors [20], who showed a sharp increase in wear coefficient when BN volume fraction was greater than 10%. The other study claimed that the BN addition has no net effect on COF and wear coefficients [168]. In this case, boron nitride is reported to be ineffective in providing a solid lubricating film at RT. On the other hand, Chen et al. [170] reported that the wear coefficient decreased sharply with increase in hBN volume fraction, for example, it was lower than 10^{-6} mm³/Nm for Si_3N_4-20% hBN.

The examples of worn surfaces of experimental materials are illustrated in Figure 4.30. Figure 4.30a and b shows the wear track of SNB5-nB material that

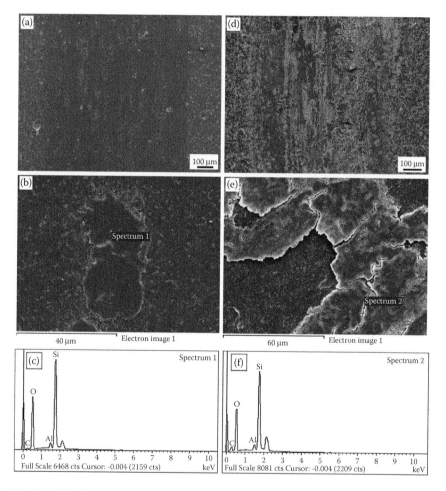

FIGURE 4.30 Worn surfaces of Si_3N_4/BN composites, (a) and (b) SNB5-nB, (c) EDX analysis of wear. Track of SNB5-nB, (d) and (e) SNB5-mE, (f) EDX analysis of weartrack of SNB5-mE. The EDX analysis clearly points out the oxygen enrichment of the debris layer. (From Kovalčíková, A. et al., *J. Euro. Ceram. Soc.*, 34, 3319–3328, 2014.)

was very small and smooth with some micro scratches. The higher amount of coherent debris has been observed in the wear tracks of micro/microcomposites, that is, in materials with higher wear resistance (SNB5-mE, Figure 4.30d and e). The main wear mechanism in this work was similar for all studied materials in the form of mechanical failure (microfracture) and tribochemical reaction. The tribochemical reactions create a film on the surface of materials and above the critical load; the tribochemical film was partially removed, resulting in microfracture in discrete regions. This tribofilm should protect the wear surface. EDX analysis of wear tracks (Figure 4.30c and f) indicated that the coherent layers contain large amount of oxygen. These tribochemical reactions can be because of the effects of humidity according to which silicon nitride and silicon carbide react with oxygen from air to form hydrate SiO_2 layer [172,173]. Moreover, due to possible high contact temperatures, also oxidation could take place. However, the oxygen content of the debris layer should be mainly attributed to humidity-driven triboreaction more than an oxidation. Skopp et al. concluded that the addition of BN to Si_3N_4 is only tribologically effective below 100°C in humid air because a film formation of BN or $BN(H_2O)_x$ on the wear surface. Furthermore, Erdemir et al. [174] ascribed very low friction coefficients to the formation of self-lubricating boric acid films (H_3BO_3) on boron-containing surfaces. As temperature increases, the lubricant layers of $BN(H_2O)_x$ and (H_3BO_3) are destroyed either by vaporization of water, or by thermal decomposition above 150°C. The decrease of COF is attributed to BN hydration, which provides an *in situ* lubrication. Gangopadhyay et al. [6] reported the ineffectiveness of BN to lubricate either alumina or silicon nitride due to the limited formation of transfer films. In this study, the formation of hydrated layers on the worn surfaces was not identified and probably therefore the COF on prepared composites are similar to that of the monolithic silicon nitride. However, the positive effect of BN on friction is accompanied by improved wear resistance, when low amounts of BN (5 wt.%) were added to silicon nitride matrix. This improvement can be related to higher fracture toughness of Si_3N_4/BN micro/ microcomposites compared with monolithic silicon nitride. In conclusion, for used weight fractions (1%, 3%, or 5%) of all types of BN additives, no decrease of coefficient of friction was observed. BN phase did not participate in lubricating process. However, introduction of boron nitride did lead to better wear resistance. The best results were found for the materials with 1 wt.% microsized BN addition whose specific wear rate was 78% lower than that of monolithic silicon nitride. The main wear mechanism is similar for all studied materials in the form of mechanical wear (microfracture) and tribochemical reaction.

4.6   ZIRCONIA-MATRIX COMPOSITES

Tetragonal zirconia polycrystals stabilized by yttria present a good combination of fracture toughness and bending strength, which is related to the stress-induced phase transformation of tetragonal ZrO_2 (Y_2O_3) into monoclinic symmetry. Therefore, zirconia ceramics are potential candidates for a host of engineering applications, especially at high temperatures. However, the friction coefficient of zirconia ceramics in dry sliding is high enough not to acceptable for engineering

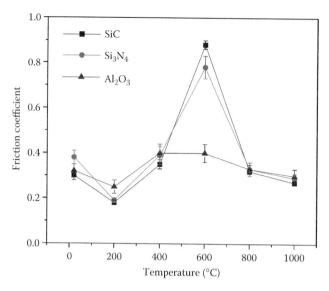

FIGURE 4.31 CoFsofZrO$_2$ (Y$_2$O$_3$) matrix composites at different tested temperatures (0.20 m/s, 10 N, 30 min). (From Kong, L. et al., *Tribo. Inter.*, 64, 53–62, 2013.)

applications. Consequently, it is quite necessary to research and develop ZrO$_2$ (Y$_2$O$_3$) matrix high-temperature self-lubricating composites. It was reported that the additives of graphite, MoS$_2$, BaF$_2$, CaF$_2$, Ag, Ag$_2$O, Cu$_2$O, BaCrO$_4$, BaSO$_4$, SrSO$_4$, and CaSiO$_3$ were incorporated into zirconia ceramics, respectively, to evaluate their potentials as effective solid lubricants over a wide operating temperature range [17,39,175].

A research investigated that the friction and wear behavior of a ZrO$_2$ (Y$_2$O$_3$)–MoS$_2$–CaF$_2$ composite sliding against commercial SiC, Si$_3$N$_4$, and Al$_2$O$_3$ ceramic balls was investigated from 20°C to 1000°C [176]; it can be clearly seen in Figure 4.31. In Figure 4.31, the Al$_2$O$_3$/ZMC10 tribosystem shows low CoFs from 20°C to 1000°C. It can be found that at RT, the COF of Al$_2$O$_3$/ZMC10 tribosystem is 0.32, as the temperature rises to 200°C, the COF is 0.27; at 400°C, it increases to 0.40, which is near to that of at 600°C; when the temperature rises to 800°C, the CoF drops to 0.30; and further it drops to 0.28 at 1000°C. It was found that the ZrO$_2$ (Y$_2$O$_3$) composites incorporated with SrSO$_4$ exhibited low steady-state friction coefficients of less than 0.2 and small wear rates in the order of 10^{-6} mm^3/Nm at low sliding speed from RT to 800°C. The formation, plastic deformation, and effective spreading of SrSO$_4$ lubricating film were the most important factor to reduce friction and wear rate over a wide temperature range. A ZrO$_2$-matrix high-temperature self-lubricating composite with addition of MoS$_2$ and CaF$_2$ as lubricants prepared using hot pressing method was investigated from RT to 1000°C [177]. The ZrO$_2$–MoS$_2$–CaF$_2$ composites had favorable microhardness (HV 824 ± 90) and fracture toughness (6.5 ± 1.4 MPa m$^{1/2}$), and against SiC ceramic exhibited excellent self-lubricating and antiwear properties at a wide

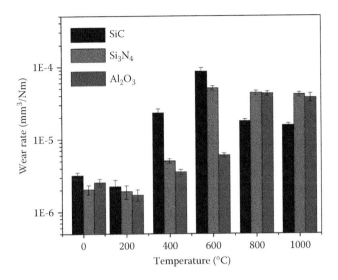

FIGURE 4.32 Variations of wear rates of the $ZrO_2–MoS_2–CaF_2$ composite at different temperatures (tested at an applied load of 10 N and sliding speed of 0.2 m/s against SiC ceramic ball). (From Kong, L. et al., *Tribo. Inter.*, 64, 53–62, 2013.)

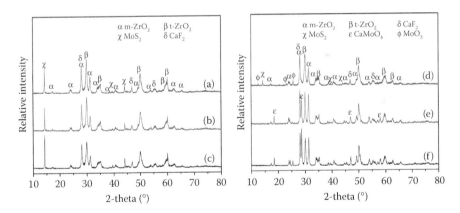

FIGURE 4.33 XRD patterns of the $ZrO_2–MoS_2–CaF_2$ composite (a) and its worn surfaces at different temperatures: 200°C (b), 400°C (c), 600°C (d), 800°C (e), and 1000°C (f). (From Kong, L. et al., *Tribo. Inter.*, 64, 53–62, 2013.)

temperature range. At 1000°C, the ZrO_2-matrix composite had a very low coefficient of friction of about 0.27 and wear rate of 1.54×10^{-5} mm³/Nm, as shown in Figure 4.32. The low friction and wear were attributed to a new lubricant $CaMoO_4$ that formed on the worn surfaces at high temperatures (Figure 4.33).

REFERENCES

1. Buckley DH, Miyoshi K. Friction and wear of ceramics. *Wear*. 1984;100(1–3):333–353.
2. Fischer TE et al. Friction and wear of tough and brittle zirconia in nitrogen, air, water, hexadecane and hexadecane containing stearic acid. *Wear*. 1988;124(2):133–148.
3. Bohmer M, Almond EA. Mechanical properties and wear resistance of a whisker-reinforced zirconia-toughened alumina. *Materials Science and Engineering: A*. 1988;105:105–116.
4. Breznak J, Breval E, Macmillan N. Sliding friction and wear of structural ceramics. *Journal of Materials Science*. 1985;20(12):4657–4680.
5. Gates R, Hsu M, Klaus E. Tribochemical mechanism of alumina with water. *Tribology Transactions*. 1989;32(3):357–363.
6. Gangopadhyay A, Jahanmir S, Peterson MB. Self-lubricating ceramic matrix composites. In: Jahanmir S, (Ed.). *Friction and Wear of Ceramics*. Marcel Dekker: New York; 1994. pp. 163–197.
7. Tomizawa H, Fischer T. Friction and wear of silicon nitride and silicon carbide in water: Hydrodynamic lubrication at low sliding speed obtained by tribochemical wear. *ASLE Transactions*. 1987;30(1):41–46.
8. Pope LE, Fehrenbacher LL, Winer WO. *New Materials Approaches to Tribology*. Materials Research Society: Pittsburgh, PA; 1989.
9. Omrani E et al. New emerging self-lubricating metal matrix composites for tribological applications. In: Davim JP, (Ed.). *Ecotribology*. Springer: Cham, Switzerland; 2016. pp. 63–103.
10. Nishimura M. Tribological problems in the space development in Japan. *JSME International Journal. Ser. 3, Vibration, Control Engineering, Engineering for Industry*. 1988;31(4):661–670.
11. Pallini R, Wedeven L. Traction characteristics of graphite lubricants at high temperature. *Tribology Transactions*. 1988;31(2):289–295.
12. Wedeven L, Pallini R, Miller N. Tribological examination of unlubricated and graphite-lubricated silicon nitride under traction stress. *Wear*, 1988;122(2):183–205.
13. Sliney H et al. Tribology of selected ceramics at temperatures to 900°C. *Tenth Annual Conference on Composites and Advanced Ceramic Materials*. American Ceramic Society: Cocoa Beach, FL; 1986.
14. Sliney HE. Wide temperature spectrum self-lubricating coatings prepared by plasma spraying. *Thin Solid Films*. 1979;64(2):211–217.
15. Liu H, Xue Q. The tribological properties of TZP-graphite self-lubricating ceramics. *Wear*. 1996;198(1):143–149.
16. Jin Y, Kato K, Umehara N. Further investigation on the tribological behavior of Al_2O_3–$20Ag_2OCaF_2$ composite at 650°C. *Tribology Letters*. 1999;6(3):225–232.
17. Ouyang J et al. Tribological properties of spark-plasma-sintered ZrO_2 (Y_2O_3)–CaF_2–Ag composites at elevated temperatures. *Wear*. 2005;258(9):1444–1454.
18. Zhang Y et al. High-performance self-lubricating ceramic composites with laminated-graded structure. In: Ebrahimi F, (Ed.). *Advances in Functionally Graded Materials and Structures*. Intech: Rijeka, Croatia; 2016. https://cdn.intechopen.com/pdfs-wm/49949.pdf
19. Zhang Y-S et al. Lubrication behavior of Y-TZP/Al_2O_3/Mo nanocomposites at high temperature. *Wear*. 2010;268(9):1091–1094.
20. Carrapichano J, Gomes J, Silva R. Tribological behaviour of Si_3N_4–BN ceramic materials for dry sliding applications. *Wear*. 2002;253(9):1070–1076.
21. Ye Y, Chen J, Zhou H. An investigation of friction and wear performances of bonded molybdenum disulfide solid film lubricants in fretting conditions. *Wear*. 2009;266(7):859–864.

22. Moshkovich A et al. Friction and wear of solid lubricant films deposited by different types of burnishing. *Wear.* 2007;263(7):1324–1327.

23. Senda T, Drennan J, McPherson R. Sliding wear of oxide ceramics at elevated temperatures. *Journal of the American Ceramic Society.* 1995;78(11):3018–3024.

24. Aizawa T et al. Self-lubrication mechanism via the in situ formed lubricious oxide tribofilm. *Wear.* 2005;259(1):708–718.

25. Wang L et al. Tribological investigation of CaF_2 nanocrystals as grease additives. *Tribology International.* 2007;40(7):1179–1185.

26. Wang Y, Liu Z. Tribological properties of high temperature self-lubrication metal ceramics with an interpenetrating network. *Wear.* 2008;265(11):1720–1726.

27. Mattern A et al. Preparation of interpenetrating ceramic–metal composites. *Journal of the European ceramic society.* 2004;24(12):3399–3408.

28. Bahadur S, Yang C-N. Friction and wear behavior of tungsten and titanium carbide coatings. *Wear.* 1996;196(1–2):156–163.

29. Van Acker K, Vercammen K. Abrasive wear by TiO_2 particles on hard and on low friction coatings. *Wear.* 2004;256(3):353–361.

30. Qi Y-E, Zhang Y-S, Hu L-T. High-temperature self-lubricated properties of Al_2O_3/Mo laminated composites. *Wear.* 2012;280:1–4.

31. Sierra C, Vazquez A. Dry sliding wear behaviour of nickel aluminides coatings produced by self-propagating high-temperature synthesis. *Intermetallics.* 2006;14(7):848–852.

32. Liu C, Pope DP. Ni_3Al and its Alloys. *Intermetallic Compounds.* 2000;2:17–47.

33. Sikka V et al. Advances in processing of Ni_3Al-based intermetallics and applications. *Intermetallics.* 2000;8(9):1329–1337.

34. Sliney H.E. Solid lubricant materials for high temperatures—A review. *Tribology International.* 1982;15(5):303–315.

35. Pawlak Z et al. A comparative study on the tribological behaviour of hexagonal boron nitride (h-BN) as lubricating micro-particles—An additive in porous sliding bearings for a car clutch. *Wear.* 2009;267(5):1198–1202.

36. Mahathanabodee S et al. Effects of hexagonal boron nitride and sintering temperature on mechanical and tribological properties of SS316L/h-BN composites. *Materials & Design.* 2013;46:588–597.

37. Zhang X-F et al. Microstructure and properties of HVOF sprayed Ni-based submicron WS_2/CaF_2 self-lubricating composite coating. *Transactions of Nonferrous Metals Society of China.* 2009;19(1):85–92.

38. Shi X et al. Tribological behavior of Ni_3Al matrix self-lubricating composites containing WS_2, Ag and hBN tested from room temperature to 800°C. *Materials & Design.* 2014;55:75–84.

39. Ouyang J et al. Spark-plasma-sintered ZrO_2 (Y_2O_3)-$BaCrO_4$ self-lubricating composites for high temperature tribological applications. *Ceramics International.* 2005;31(4):543–553.

40. Murakami T et al. High-temperature tribological properties of spark-plasma-sintered Al_2O_3 composites containing barite-type structure sulfates. *Tribology International.* 2007;40(2):246–253.

41. Gulbiński W, Suszko T. Thin films of MoO_3–Ag_2O binary oxides–the high temperature lubricants. *Wear.* 2006;261(7):867–873.

42. Zhu S et al. Barium chromate as a solid lubricant for nickel aluminum. *Tribology Transactions.* 2012;55(2):218–223.

43. Zhu S et al. Tribological property of Ni_3Al matrix composites with addition of $BaMoO_4$. *Tribology Letters.* 2011;43(1):55–63.

44. Deadmore DL, Sliney HE. Hardness of CaF_2 and BaF_2 solid lubricants at 25 to 670°C. NASA TM 88979. 1987.

45. Sliney HE, Strom TN, Allen GP. Fluoride solid lubricants for extreme temperatures and corrosive environments. *ASLE Transactions*. 1965;8(4):307–322.

46. Zhu S. Fabrication and tribological performance of Ni–Al matrix high temperature self-lubricating composites. PhD Thesis. Chinese Academy of Sciences, Lanzhou; 2011.

47. Zhang Y et al. Ti_3SiC_2—a self-lubricating ceramic. *Materials Letters*. 2002;55(5): 285–289.

48. Souchet A et al. Tribological duality of Ti_3SiC_2. *Tribology Letters*. 2005;18(3):341–352.

49. Shi X et al. Tribological behavior of Ti_3SiC_2/(WC–10Co) composites prepared by spark plasma sintering. *Materials & Design*. 2013;45:365–376.

50. Su Y-L, Kao W-H. Tribological behaviour and wear mechanism of MoS_2–Cr coatings sliding against various counterbody. *Tribology International*. 2003;36(1):11–23.

51. Yao J et al. Influence of lubricants on wear and self-lubricating mechanisms of Ni_3Al matrix self-lubricating composites. *Journal of Materials Engineering and Performance*. 2015;24(1):280–295.

52. Wu X et al. Aqueous mineralization process to synthesize uniform shuttle-like $BaMoO_4$ microcrystals at room temperature. *Journal of Solid State Chemistry*. 2007;180(11):3288–3295.

53. Basiev T et al. Spontaneous Raman spectroscopy of tungstate and molybdate crystals for Raman lasers. *Optical Materials*. 2000;15(3):205–216.

54. Gabr R, El-Awad A, Girgis M. Physico-chemical and catalytic studies on the calcination products of $BaCrO_4$–CrO_3 mixture. *Materials Chemistry and Physics*. 1992;30(4):253–259.

55. Azad A, Sudha R, Sreedharan O. The standard Gibbs energies of formation of $ACrO_4$ (A = Ca, Sr or Ba) from EMF measurements. *Thermochimica Acta*. 1992;194:129–136.

56. Guo J et al. Tensile properties and microstructures of $NiAl$-$20TiB_2$ and $NiAl$-$20TiC$ in situ composites. *Materials & Design*. 1997;18(4):357–360.

57. Bhaumik S et al. Reaction sintering of NiAl and TiB_2–NiAl composites under pressure. *Materials Science and Engineering: A*. 1998;257(2):341–348.

58. Gao M et al. Interpenetrating microstructure and fracture mechanism of NiAl/TiC composites by pressureless melt infiltration. *Materials Letters*. 2004;58(11):1761–1765.

59. Alman DE, Hawk JA. The abrasive wear of sintered titanium matrix–ceramic particle reinforced composites. *Wear*. 1999;225:629–639.

60. Johnson B, Kennedy F, Baker I. Dry sliding wear of NiAl. *Wear*. 1996;192(1–2):241–247.

61. Jin J-H, Stephenson D. The sliding wear behaviour of reactively hot pressed nickel aluminides. *Wear*. 1998;217(2):200–207.

62. Zhu S et al. Tribological behavior of NiAl matrix composites with addition of oxides at high temperatures. *Wear*. 2012;274:423–434.

63. Shi X et al. Influence of Ti_3SiC_2 content on tribological properties of NiAl matrix self-lubricating composites. *Materials & Design*. 2013;45:179–189.

64. Ozdemir O, Zeytin S, Bindal C. Tribological properties of NiAl produced by pressure-assisted combustion synthesis. *Wear*. 2008;265(7):979–985.

65. Centers PW. The role of oxide and sulfide additions in solid lubricant compacts. *Tribology Transactions*. 1988;31(2):149–156.

66. Ohno N et al. Tribological properties and film formation behavior of thermoreversible gel lubricants. *Tribology Transactions*. 2010;53(5):722–730.

67. Kondo H, Aoki M, Seto JE. Tribochemical reactions on the surfaces of thin film magnetic media. *Tribology Transactions*. 1993;36(2):193–200.

68. Johnson D et al. Processing and mechanical properties of in-situ composites from the NiAlCr and the NiAl (Cr, Mo) eutectic systems. *Intermetallics*. 1995;3(2):99–113.

69. Huai K et al. Microstructure and mechanical behavior of NiAl-based alloy prepared by powder metallurgical route. *Intermetallics*. 2007;15(5):749–752.

70. Zhu S et al. NiAl matrix high-temperature self-lubricating composite. *Tribology Letters*. 2011;41(3):535–540.

71. Hu KH, Hu XG, Sun XJ. Morphological effect of MoS_2 nanoparticles on catalytic oxidation and vacuum lubrication. *Applied Surface Science*. 2010;256(8):2517–2523.

72. Arslan E et al. High temperature friction and wear behavior of MoS_2/Nb coating in ambient air. *Journal of Coatings Technology and Research*. 2010;7(1):131.

73. Zabinski J et al. Chemical and tribological characterization of PbO MoS_2 films grown by pulsed laser deposition. *Thin Solid Films*. 1992;214(2):156–163.

74. Sen R et al. Encapsulated and hollow closed-cage structures of WS_2 and MoS_2 prepared by laser ablation at 450–1050°C. *Chemical Physics Letters*. 2001;340(3):242–248.

75. Pampuch R et al. Solid combustion synthesis of Ti_3SiC_2. *Journal of the European Ceramic Society*. 1989;5(5):283–287.

76. Finkel P, Barsoum M, El-Raghy T. Low temperature dependence of the elastic properties of Ti_3SiC_2. *Journal of Applied Physics*. 1999;85(10):7123–7126.

77. Finkel P, Barsoum M, El-Raghy T. Low temperature dependencies of the elastic properties of Ti_4AlN_3, $Ti_3Al_{1.1}C_{1.8}$, and Ti_3SiC_2. *Journal of Applied Physics*. 2000;87(4):1701–1703.

78. Shi X et al. Synergetic lubricating effect of MoS_2 and Ti_3SiC_2 on tribological properties of NiAl matrix self-lubricating composites over a wide temperature range. *Materials & Design*. 2014;55:93–103.

79. Kong L et al. High-temperature tribological behavior of ZrO_2–MoS_2–CaF_2 self-lubricating composites. *Journal of the European Ceramic Society*. 2013;33(1):51–59.

80. Wong K et al. Surface and friction characterization of MoS_2 and WS_2 third body thin films under simulated wheel/rail rolling–sliding contact. *Wear*. 2008;264(7):526–534.

81. Shi X et al. Tribological behaviors of NiAl based self-lubricating composites containing different solid lubricants at elevated temperatures. *Wear*. 2014;310(1):1–11.

82. Lee C et al. Measurement of the elastic properties and intrinsic strength of monolayer graphene. *Science*. 2008;321(5887):385–388.

83. Soldano C, Mahmood A, Dujardin E. Production, properties and potential of graphene. *Carbon*. 2010;48(8):2127–2150.

84. Berman D, Erdemir A, Sumant AV. Reduced wear and friction enabled by graphene layers on sliding steel surfaces in dry nitrogen. *Carbon*. 2013;59:167–175.

85. Xiao Y et al. Tribological performance of NiAl self-lubricating matrix composite with addition of graphene at different loads. *Journal of Materials Engineering and Performance*. 2015;24(8):2866–2874.

86. Bhushan B. *Nanotribology and Nanomechanics I: Measurement Techniques and Nanomechanics*, Vol. 1. Springer Science & Business Media: Berlin, Germany; 2011.

87. Bhushan B. Nanotribology, nanomechanics, and materials characterization. In: Bhushan B, (Ed.). *Springer Handbook of Nanotechnology*. Springer: Berlin, Germany; 2010. pp. 789–856.

88. Qian L et al. Nanofretting behaviors of NiTi shape memory alloy. *Wear*. 2007;263(1):501–507.

89. Bhushan B. Nanotribology and nanomechanics. *Wear*. 2005;259(7):1507–1531.

90. Schaupp D, Schneider J, Zum GK-H. Wear mechanisms on multiphase Al_2O_3 ceramics during running-in period in unlubricated oscillating sliding contact. *Tribology Letters*. 2001;9(3):125–131.

91. Ziebert C, Gahr K-HZ. Microtribological properties of two-phase Al_2O_3 ceramic studied by AFM and FFM in air of different relative humidity. *Tribology Letters*. 2004;17(4):901–909.

92. Novak S, Kalin M. The effect of pH on the wear of water-lubricated alumina and zirconia ceramics. *Tribology Letters*. 2004;17(4):727–732.

93. Zhou Z et al. Friction and wear properties of ZrO_2–Al_2O_3 composite with three layered structure under water lubrication. *Tribology Letters*. 2013;49(1):151–156.

94. Jin Y, Kato K, Umehara N. Tribological properties of self-lubricating CMC/Al_2O_3 pairs at high temperature in air. *Tribology Letters*. 1998;4(3):243–250.

95. Jin Y, Kato K, Umehara, N. Effects of sintering aids and solid lubricants on tribological behaviours of CMC/Al_2O_3 pair at 650°C. *Tribology Letters*. 1999;6(1):15–21.

96. Hsiao W et al. Wear resistance and microstructural properties of Ni–Al/h-BN/WC–Co coatings deposited using plasma spraying. *Materials Characterization*. 2013;79:84–92.

97. Zishan C et al. Tribological behaviors of SiC/h-BN composite coating at elevated temperatures. *Tribology International*. 2012;56:58–65.

98. Li X et al. Fabrication and characterization of B_4C-based ceramic composites with different mass fractions of hexagonal boron nitride. *Ceramics International*. 2015;41(1):27–36.

99. Kovalčíková A et al. Influence of hBN content on mechanical and tribological properties of Si_3N_4/BN ceramic composites. *Journal of the European Ceramic Society*. 2014;34(14):3319–3328.

100. Cho M, Kim D, Cho W. Analysis of micro-machining characteristics of Si_3N_4–hBN composites. *Journal of the European Ceramic Society*. 2007;27(2):1259–1265.

101. Yuan B et al. Silicon nitride/boron nitride ceramic composites fabricated by reactive pressureless sintering. *Ceramics International*. 2009;35(6):2155–2159.

102. Blau P et al. Reciprocating friction and wear behavior of a ceramic-matrix graphite composite for possible use in diesel engine valve guides. *Wear*. 1999;225:1338–1349.

103. Dong L-M et al. A study on the friction and wear behavior of ceramic-graphite composite. *Tribology-Beijing-*. 1997;17:361–366.

104. Qi Y et al. Design and preparation of high-performance alumina functional graded self-lubricated ceramic composites. *Composites Part B: Engineering*. 2013;47:145–149.

105. Song J et al. Influence of structural parameters and compositions on the tribological properties of alumina/graphite laminated composites. *Wear*. 2015;338:351–361.

106. Wang C-A et al. Biomimetic structure design—A possible approach to change the brittleness of ceramics in nature. *Materials Science and Engineering: C*. 2000;11(1):9–12.

107. Mekky W, Nicholson PS. The fracture toughness of Ni/Al_2O_3 laminates by digital image correlation I: Experimental crack opening displacement and R-curves. *Engineering Fracture Mechanics*. 2006;73(5):571–582.

108. Mekky W, Nicholson PS. The fracture toughness of Ni/Al_2O_3 laminates by digital image correlation II: Bridging-stresses and R-curve models. *Engineering Fracture Mechanics*. 2006;73(5):583–592.

109. Launey ME et al. A novel biomimetic approach to the design of high-performance ceramic–metal composites. *Journal of the Royal Society Interface*. 2010;7(46):741–753.

110. Song J et al. Influence of structural parameters and transition interface on the fracture property of Al_2O_3/Mo laminated composites. *Journal of the European Ceramic Society*. 2015;35(5):1581–1591.

111. Fang Y et al. Influence of structural parameters on the tribological properties of Al_2O_3/Mo laminated nanocomposites. *Wear*. 2014;320:152–160.

112. Su Y et al. High-temperature self-lubricated and fracture properties of alumina/molybdenum fibrous monolithic ceramic. *Tribology Letters*. 2016;61(1):9.

113. Fang Y et al. Design and fabrication of laminated–graded zirconia self-lubricating composites. *Materials & Design*. 2013;49:421–425.

114. Schwartz MM. *Composite Materials. Volume 1: Properties, Non-Destructive Testing, and Repair*. Prentice Hall: Old Tappan, NJ; 1997.

115. Yang X, Rahaman M. SiC platelet-reinforced Al_2O_3 composites by free sintering of coated inclusions. *Journal of the European Ceramic Society*. 1996;16(11):1213–1220.

116. Hu CL, Rahaman MN. SiC-whisker-reinforced Al_2O_3 composites by free sintering of coated powders. *Journal of the American Ceramic Society*. 1993;76(10):2549–2554.

117. Koichi N. New design concept of structural ceramics-ceramics nanocomposites. *Journal of Ceramic Society of Japan*. 1991;99(3):974–82.

118. Lawn B, Fuller E. New design concept of structural ceramics: Ceramic nanocomposites. *Journal of Material Science*. 1975;12:2016–2024.

119. Deb A, Chatterjee P, Gupta SS. Synthesis and microstructural characterization of α-Al_2O_3-t-ZrO_2 composite powders prepared by combustion technique. *Materials Science and Engineering: A*. 2007;459(1):124–131.

120. Ruehle M, Claussen N, Heuer AH. Transformation and microcrack toughening as complementary processes in ZrO_2-toughened Al_2O_3. *Journal of the American Ceramic Society*. 1986;69(3):195–197.

121. Kim S-H, Lee SW. Wear and friction behavior of self-lubricating alumina–zirconia–fluoride composites fabricated by the PECS technique. *Ceramics International*. 2014;40(1):779–790.

122. Yang X-F et al. Wear properties and microstructures of alumina matrix composite ceramics used for drawing dies. *Ceramics International*. 2009;35(8):3495–3502.

123. Jianxin D, Xuefeng Y, Jinghai W. Wear mechanisms of Al_2O_3/TiC/Mo/Ni ceramic wire-drawing dies. *Materials Science and Engineering: A*. 2006;424(1):347–354.

124. Song P et al. Tribological properties of self-lubricating laminated ceramic materials. *Journal of Wuhan University of Technology. Materials Science Edition*. 2014;29(5):906.

125. Zhang X-Y, Tan S-H, Jiang D-L. AlN–TiB_2 composites fabricated by spark plasma sintering. *Ceramics International*. 2005;31(2):267–270.

126. Kawabata T, Fukai H, Izumi O. Effect of ternary additions on mechanical properties of TiAl. *Acta Materialia*. 1998;46(6):2185–2194.

127. Zollinger J et al. Influence of oxygen on solidification behaviour of cast TiAl-based alloys. *Intermetallics*. 2007;15(10):1343–1350.

128. Wu X. Review of alloy and process development of TiAl alloys. *Intermetallics*. 2006;14(10):1114–1122.

129. Appel F, Wagner R. Microstructure and deformation of two-phase γ-titanium aluminides. *Materials Science and Engineering: R: Reports*. 1998;22(5):187–268.

130. Rastkar A, Shokri B. A multi-step process of oxygen diffusion to improve the wear performance of a gamma-based titanium aluminide. *Wear*. 2008;264(11):973–979.

131. Yuan Y et al. The precipitation reaction in a Ag-modified TiAl based intermetallic, as studied by TEM. *Journal of Alloys and Compounds*. 2005;399(1):126–131.

132. Shu S et al. Compression properties and work-hardening behavior of Ti_2 AlC/TiAl composites fabricated by combustion synthesis and hot press consolidation in the Ti–Al–Nb–C system. *Materials & Design*. 2011;32(10):5061–5065.

133. Imayev R et al. Alloy design concepts for refined gamma titanium aluminide based alloys. *Intermetallics*. 2007;15(4):451–460.

134. Li C, Xia J, Dong H. Sliding wear of TiAl intermetallics against steel and ceramics of Al_2O_3, Si_3N_4 and WC/Co. *Wear*. 2006;261(5):693–701.

135. Cheng J et al. Effect of TiB_2 on dry-sliding tribological properties of TiAl intermetallics. *Tribology International*. 2013;62:91–99.

136. Liu X-B, Yu R-L. Influences of precursor constitution and processing speed on microstructure and wear behavior during laser clad composite coatings on γ-TiAl intermetallic alloy. *Materials & Design*. 2009;30(2):391–397.

137. Sun T et al. Study on dry sliding friction and wear properties of Ti_2 AlN/TiAl composite. *Wear*. 2010;268(5):693–699.

138. Cheng J et al. The tribological behavior of a Ti-46Al-2Cr-2Nb alloy under liquid paraffine lubrication. *Tribology Letters*. 2012;46(3):233–241.

139. Miyoshi K, Lerch BA, Draper SL. Fretting wear of Ti-48Al-2Cr-2Nb. *Tribology International*. 2003;36(2):145–153.

140. Cheng J et al. High temperature tribological behavior of a Ti-46Al-2Cr-2Nb intermetallics. *Intermetallics*. 2012;31:120–126.

141. Zhai H et al. Oxidation layer in sliding friction surface of high-purity Ti_3SiC_2. *Journal of Materials Science*. 2004;39(21):6635–6637.

142. Zhai H, Huang Z, Ai M. Tribological behaviors of bulk Ti_3SiC_2 and influences of TiC impurities. *Materials Science and Engineering: A*. 2006;435:360–370.

143. Li J-F et al. Mechanical properties of polycrystalline Ti_3SiC_2 at ambient and elevated temperatures. *Acta Materialia*. 2001;49(6):937–945.

144. Radovic M et al. Effect of temperature, strain rate and grain size on the mechanical response of Ti_3SiC_2 in tension. *Acta Materialia*. 2002;50(6):1297–1306.

145. Kooi B et al. Ti_3SiC_2: A damage tolerant ceramic studied with nano-indentations and transmission electron microscopy. *Acta Materialia*. 2003;51(10):2859–2872.

146. Zhai HX et al. Frictional layer and its antifriction effect in high-purity Ti_3SiC_2 and TiC-contained Ti_3SiC_2. *Key Engineering Materials*. 2005;280–283:1347–1352.

147. Huang ZY et al. Sliding friction behavior of bulk Ti_3SiC_2 under difference normal pressures. *Key Engineering Materials*. 2005;280–283:1353–1356.

148. Zhang ZL et al. Tribo-chemical reaction in bulk Ti_3SiC_2 under sliding friction. *Key Engineering Materials*. 2005;280–283:1357–1360.

149. Barsoum MW, El-Raghy T, Ogbuji LU. Oxidation of Ti_3SiC_2 in air. *Journal of the Electrochemical Society*. 1997;144(7):2508–2516.

150. Gupta S et al. Tribological behavior of select MAX phases against Al_2O_3 at elevated temperatures. *Wear*. 2008;265(3):560–565.

151. Xu Z et al. High-temperature tribological performance of Ti_3SiC_2/TiAl self-lubricating composite against Si_3N_4 in Air. *Journal of Materials Engineering and Performance*. 2014;23(6):2255–2264.

152. Kim G et al. Characterization of atmospheric plasma spray NiCr–Cr_2O_3–Ag–CaF_2/BaF_2 coatings. *Surface and Coatings Technology*. 2005;195(1):107–115.

153. Yuan J et al. Microstructures and tribological properties of plasma sprayed WC–Co–Cu–BaF_2/CaF_2 self-lubricating wear resistant coatings. *Applied Surface Science*. 2010;256(16):4938–4944.

154. Shi X et al. Tribological performance of TiAl matrix self-lubricating composites containing Ag, Ti_3SiC_2 and BaF_2/CaF_2 tested from room temperature to 600°C. *Materials & Design*. 2014;53:620–633.

155. Riley FL. Silicon nitride and related materials. *Journal of the American Ceramic Society*. 2000;83(2):245–265.

156. Skopp A, Woydt M, Habig K-H. Tribological behavior of silicon nitride materials under unlubricated sliding between 22°C and 1000°C. *Wear*. 1995;181:571–580.

157. Strong KL, Zabinski JS. Tribology of pulsed laser deposited thin films of cesium oxythiomolybdate (CS_2MoOS_3). *Thin Solid Films*. 2002;406(1):174–184.

158. Strong KL, Zabinski JS. Characterization of annealed pulsed laser deposited (PLD) thin films of cesium oxythiomolybdate (CS_2MoOS_3). *Thin Solid Films*. 2002;406(1):164–173.

159. Rosado L et al. Solid lubrication of silicon nitride with cesium-based compounds: Part I—rolling contact endurance, friction and wear. *Tribology Transactions*. 2000;43(3):489–497.

160. Rosado L, Forster NH, Wittberg TN. Solid lubrication of silicon nitride with cesium-based compounds: Part II—surface analysis. *Tribology Transactions*. 2000;43(3):521–527.

161. Gao L et al. BN/Si_3N_4 nanocomposite with high strength and good machinability. *Materials Science and Engineering: A*. 2006;415(1):145–148.

162. Wei D, Meng Q, Jia D. Mechanical and tribological properties of hot-pressed h-BN/Si$_3$N$_4$ ceramic composites. *Ceramics International*. 2006;32(5):549–554.
163. Skopp A, Woydt M. Ceramic-ceramic composite materials with improved friction and wear properties. *Tribology International*. 1992;25(1):61–70.
164. Ruigang W et al. Investigation of the physical and mechanical properties of hot-pressed machinable Si$_3$N$_4$/h-BN composites and FGM. *Materials Science and Engineering: B*. 2002;90(3):261–268.
165. Sun Y et al. Effect of hexagonal BN on the microstructure and mechanical properties of Si$_3$N$_4$ ceramics. *Journal of Materials Processing Technology*. 2007;182(1):134–138.
166. Li Y-L, Li R-X, Zhang J-X. Enhanced mechanical properties of machinable Si$_3$N$_4$/BN composites by spark plasma sintering. *Materials Science and Engineering: A*. 2008;483:207–210.
167. Liu H, Hsu SM. Fracture behavior of multilayer silicon nitride/boron nitride ceramics. *Journal of the American Ceramic Society*. 1996;79(9):2452–2457.
168. Saito T, Hosoe T, Honda F. Chemical wear of sintered Si$_3$N$_4$, hBN and Si$_3$N$_4$–hBN composites by water lubrication. *Wear*. 2001;247(2):223–230.
169. Larsson P, Axen N, Hogmark S. Tribofilm formation on boron carbide in sliding wear. *Wear*. 1999;236(1):73–80.
170. Chen W et al. Tribological characteristics of Si$_3$N$_4$–hBN ceramic materials sliding against stainless steel without lubrication. *Wear*. 2010;269(3):241–248.
171. Iwasa M, Kakiuchi S. Mechanical and tribological properties of Si$_3$N$_4$–BN composite ceramics. *Journal of the Ceramic Society of Japan*. 1985;93(10):661.
172. Jones MI et al. Effect of rare-earth species on the wear properties of α sialon and β silicon nitride ceramics under tribochemical type conditions. *Journal of Materials Research*. 2004;19(09):2750–2758.
173. Zum Gahr K-H et al. Micro-and macro-tribological properties of SiC ceramics in sliding contact. *Wear*. 2001;250(1):299–310.
174. Erdemir A, Bindal C. Formation and self-lubricating mechanisms of boric acid on borided steel surfaces. *Surface & Coatings Technology*. 1995;76(1–3):443–449.
175. Ouyang J, Sasaki S, Umeda K. Microstructure and tribological properties of low-pressure plasma-sprayed ZrO$_2$–CaF$_2$–Ag$_2$O composite coating at elevated temperature. *Wear*. 2001;249(5):440–451.
176. Kong L et al. ZrO$_2$ (Y$_2$O$_3$)–MoS$_2$–CaF$_2$ self-lubricating composite coupled with different ceramics from 20°C to 1000°C. *Tribology International*. 2013;64:53–62.
177. Kong L et al. Effect of CuO on self-lubricating properties of ZrO$_2$ (Y$_2$O$_3$)–Mo composites at high temperatures. *Journal of the European Ceramic Society*. 2014;34(5):1289–1296.

5 Computational Methods of Tribology in Self-Lubricating Materials

5.1 INTRODUCTION TO MOLECULAR DYNAMICS

Molecular dynamics is a technique to calculate the motion of particles in solids, liquids, and gases. Particles are atoms or molecules, and the word *motion* consists of position, velocity, and orientation that change with time for individual particles in many body systems. Interactions between particles in the system as well as system–environment interaction can be taken into account. Molecular dynamics simulation is a useful technique for micrometer focus of mesoscale modeling as presented in Figure 5.1. Molecular dynamics can be divided into two groups: first-principles molecular dynamics (FPMD) and classical molecular dynamics. The FPMD simulation relies on the fundamental properties and laws of quantum mechanics, and it does not require any empirical assumptions. Though the FPMD simulation is highly accurate, but for larger system, the computational cost makes it impossible to implement. Alder and his colleagues developed the classical molecular dynamics, and the first paper came out in 1957 [1]. In classical molecular dynamics simulation, the position and velocity of particles are calculated by solving Newton's equation. This simulation technique allows predicting the structure and properties of systems much larger in terms of the number of particles than the FPMD simulation.

The steps used in molecular dynamics simulation are similar to those in any time-dependent experiment. The first step is the preparation of sample, and selection and initialization of an N-body system by providing the initial position and velocities to an individual particle. The second step is the determination of interactions between particles with an appropriate boundary condition. The interactions between particles can be determined by calculating the force on each particle due to its all neighbor particles, and it is described by potential energy function (e.g., Lennard–Jones potential). Generally, the potential energy function ($U(\vec{r}^N)$) is generated by pairwise addition of all interactions:

$$U(\vec{r}^N) = \sum\sum u(r_{i,j}), \ i < j \tag{5.1}$$

where:
$\vec{r}^N$ is the set of vectors of all particles
$u(r_{i,j})$ is the pair potential function
$r_{i,j}$ is the distance between particles i and j

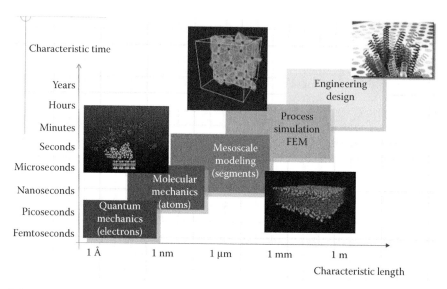

FIGURE 5.1 Example of the different times and space scales in the field of the materials modeling: quantum mechanics, molecular mechanics, molecular dynamics (mesoscale modeling), and finite element method. (From Multiscale Molecular Modeling, 2017.)

Force ($\vec{F}_i$) on each particle is determined by the potential energy function:

$$\vec{F}_i = -\nabla U(\vec{r}^N) \tag{5.2}$$

After force calculation between particles, the third step is to integrate the Newton's equation of motion (Equation 5.3) until the desired length of time is reached. For equilibrium studies, the integration continues until we do not observe any changes in the properties of the system with time:

$$\vec{F}_i(t) = m\ddot{\vec{r}}_i(t) \tag{5.3}$$

where:

    m is mass of particle
    $\ddot{\vec{r}}_i(t)$ is the acceleration

The second and third steps are the core calculation of molecular dynamics simulation. Various algorithms based on finite difference method have been developed to integrate Equation 5.3: Verlet's algorithm, general predictor–corrector algorithm, Gear predictor–corrector algorithm, leap-frog algorithm, velocity Verlet and Beeman's algorithm. We take Verlet's algorithm for explaining the integration process, and it is the combination of two Talyer's series expansion—one forward and one backward in time:

$$r_i(t+\Delta t) = r_i(t) + \frac{dr_i(t)}{dt}\Delta t + \frac{1}{2}\frac{F_i(t)}{m_i}\Delta t^2 + \frac{1}{6}\frac{d^3 r_i(t)}{dt^3}\Delta t^3 + O(\Delta t^4) \tag{5.4}$$

$$r_i(t + \Delta t) = r_i(t) - \frac{dr_i(t)}{dt}\Delta t + \frac{1}{2}\frac{F_i(t)}{m_i}\Delta t^2 - \frac{1}{6}\frac{d^3 r_i(t)}{dt^3}\Delta t^3 + O(\Delta t^4) \qquad (5.5)$$

By adding Equations 5.4 and 5.5, we have

$$r_i(t + \Delta t) = 2r_i(t) - r_i(t - \Delta t) + \frac{F_i(t)}{m_i}\Delta t^2 + O(\Delta t^4) \qquad (5.6)$$

It has a truncation error of $O(\Delta t^4)$ in each time step, and it is the third order. In this algorithm, velocities (v_i) are estimated by the first-order central difference:

$$v_i(t) = \frac{r_i(t + \Delta t) - r_i(t + \Delta t)}{2\Delta t} \qquad (5.7)$$

It can be seen that Verlet's algorithm is a two-step method with one forward and one backward time calculation. It also implies that the initial position and velocity are not sufficient to start the simulation, so we need to have a special condition to get $r_i(t - \Delta t)$. Finally, Equations 5.6 and 5.7 are solved in an iterative fashion for desired period of time and result in position and velocity of particles, respectively.

Nonequilibrium studies have also been carried out by molecular dynamic simulation. These studies were first conducted in the early 1970s [3–5]. For nonequilibrium studies, an external force is applied to the system, and then the system's response to the external force is computed. It has been used to calculate the viscosity, thermal conductivity, friction, and diffusion coefficients [6,7].

Various software packages are available for molecular dynamics simulations such as GROMACS [8], NAMD [9], LAMMPS [10], and MedeA [11]. Each software has its pros and cons, and the selection of software is based on the criteria such as computation efficiency, implementation of stable numerical integral methods for solving Newton's equation, and using different ensemble (e.g., NPT, NVT) and boundary conditions.

5.2 FRICTIONAL STUDIES AND MOLECULAR DYNAMICS

Atomic force microscope (AFM) has been used to measure frictional parameters on length scales from nanometers to micrometers in various studies [12–17]. In AFM, a sharp nanoscale tip is fixed at the free end of an oscillating cantilever, which moves on the surface of the sample with a small amount of normal load. The deflection and motion of cantilever are recorded using a detector and a friction force image is produced. An AFM image of graphene on SiC substrate is shown in Figure 5.2a, and due to adhesion between the tip and graphene, the stick-slip pattern can be seen in the frictional force map (Figure 5.2b) [17]. These nano- or microscale frictional phenomena can also be predicted and modeled by molecular dynamics simulations.

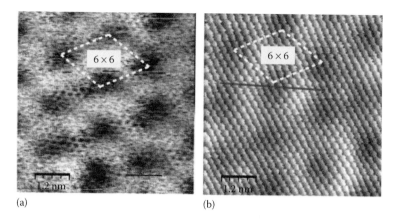

(a)   (b)

FIGURE 5.2 Graphene film on SiC substrate: (a) atomically resolved noncontact mode AFM image and (b) frictional force map. A corrugation with a 6 × 6 periodicity of the underlying SiC lattice. (From Filleter, T. and Bennewitz, R., *Phys. Rev. B*, 81, 155412, 2010.)

Mulliah et al. modeled the stick-slip phenomena of a pyramidal diamond tip into Ag (101) surface [18]. In molecular dynamics simulation, the diamond tip was simulated such that <111> direction of the diamond lattice became normal to the Ag (101) surface. Interactions between diamond–diamond, Ag–Ag, and diamong–Ag atoms were modeled by Brenner C–C potential, Ackland embedded atom method potential, and Ziegler–Biersack–Littmark potential, respectively. The simulation was carried out at various sliding speeds with an indentation depth of 5 Å. The stick-slip phenomena can be clearly observed in Figure 5.3 as the indenter's displacement is getting ups and downs after 4 ns. Apart from the stick-slip phenomena,

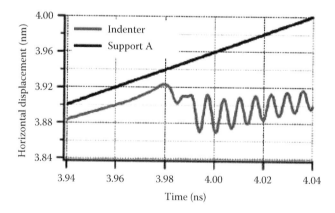

FIGURE 5.3 Displacement of indenter (diamond) and support as a function of time on Ag (101) surface with a sliding velocity of 1 m/s. (From Mulliah, D. et al., *Phys. Rev. B*, 69, 205407, 2004.)

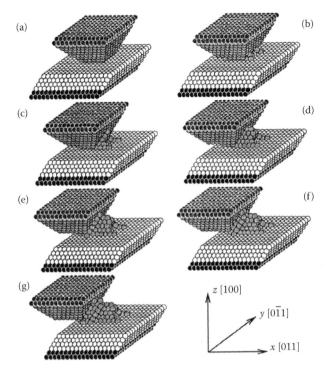

FIGURE 5.4   Successive deformation and transfer of Cu tip to the Cu substrate. Part (a) shows that initial arrangement of atoms and part (b) to (g) show deformation after 2, 4, 6, 8, 10, and 12 slips at 0 K. (From Sørensen, M.R. et al., *Phys. Rev. B Cond. Mat.*, 53, 2101–2113, 1996.)

the transfer of material from one surface to another can also be modeled using molecular dynamics simulation. One such model was reported by Sørensen et al. [19]. The authors generated a flat tip of Cu (111) over the Cu (111) surface. A fixed boundary condition was used in the direction normal to the surface, and in the other two directions, periodic boundary condition was used. The interatomic potential was derived from the effective medium theory for Cu–Cu interaction. The snapshots of the deformation of Cu tip are presented in Figure 5.4. The same behavior was also predicted using molecular dynamics simulation at 12 and 300 K with a sliding velocity of 2 m/s.

Various frictional studies have been carried out using molecular dynamics such as different deformation regimes on Cu by sliding diamond tip [20]; ploughing friction on Ag, Fe, and Si [21,22]; friction and microstructural evaluation on Ni [23]; friction force anisotropy between Si tip and substrate [24]; multicycle loading on single asperity of Au by Au flat surface [25]; the effect of stiffness on frictional properties of graphene [26]; and the influence of temperature and roughness on dry sliding [27]. These studies suggested that molecular dynamics simulation is a promising technique to predict and model the frictional studies from the atomic level to the micro level.

5.3 MOLECULAR DYNAMICS FOR SELF-LUBRICATING COMPOSITES

In self-lubricating composites, solid lubricants are dispersed uniformly through-out the component, thus eliminating the need of external lubricants. It has a wide application in aerospace and high-temperature applications such as gas turbine components. A good range of self-lubricating composites is available in the market, which includes PTFE-based metal-polymer, thermoplastic-based metal-polymer, graphite-metal, and MoS_2-based composites. One of the self-lubricating bearings is shown in Figure 5.5, which has a MoS_2-based composite cage.

5.3.1 Wear Mechanism in Self-Lubricating Materials

To model the self-lubricating materials using molecular dynamics simulation, it is highly important to understand the wear mechanism in this class of materials. We consider some examples of self-lubricating material studies. Hu et al. studied the tribological properties of MoS_2-polyoxymethylene-based self-lubricating materials [29]. The composites were prepared by blending the MoS_2 nano-balls with the polyoxymethylene at 185°C, and the tribological test was carried out by rubbing against ASTM 1045 steel with a rotating speed of 0.8 m/s under 480 N load. The composite material with 1 wt.% MoS_2 had found to have ~22% lesser coefficient of friction than the pure polyoxymethylene. The wear scar of pure polyoxymethylene and composites is shown in Figures 5.6 and 5.7, respectively. Figure 5.6 indicates the heavily worn surface of pure polyoxymethylene, and the authors also suggested that frictional heat caused the melting of the polymer under the testing condition and it led to this worn surface. The addition of MoS_2 resulted in the smooth worn surfaces (Figure 5.7) and also lower wear rate.

Tabandeh-Khorshid et al. studied the graphene-aluminum-based self-lubricating materials synthesized by powder metallurgy route [30]. They reported that Al-0.1 wt.% graphene composites had lesser worn surface after tribological tests at 5 N with 100 rpm. The worn surfaces are shown in Figure 5.8. However, with the increased amount of graphene, that is, Al-1 wt.% graphene composites had a significant num-ber of grooves on the worn surfaces and higher wear rate. This suggests that it is

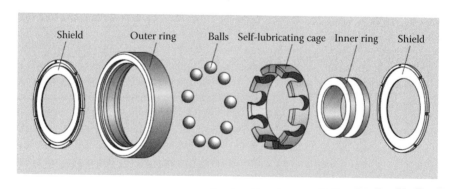

FIGURE 5.5 MoS_2-based self-lubricating ball bearing. (From Bearings for vacuum envi-ronments, products, NSK Global, 2017.)

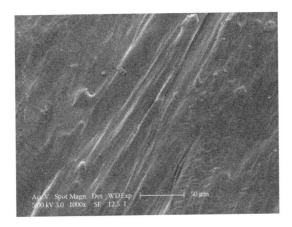

FIGURE 5.6 Scanning electron microscope (SEM) micrograph for wear scar of pure poly-oxymethylene. (From Hu, K.H. et al., *Wear*, 266, 1198–1207, 2009.)

(a)　　　　　　　　　　(b)

FIGURE 5.7 SEM micrograph for wear scar of polyoxymethylene-MoS$_2$ composites: (a) 0.5% and (b) 1 wt.% MoS$_2$. (From Hu, K.H. et al., *Wear*, 266, 1198–1207, 2009.)

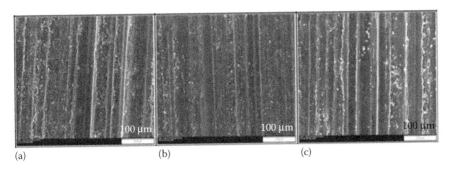

(a)　　　　　　(b)　　　　　　(c)

FIGURE 5.8 SEM micrograph of the worn surface for (a) pure Al, (b) Al-0.1 wt.% graphene, and (c) Al-1 wt.% grapheme. (From Tabandeh-Khorshid, M. et al., *Eng. Sci. Technol. Inter. J.*, 19, 463–469, 2016.)

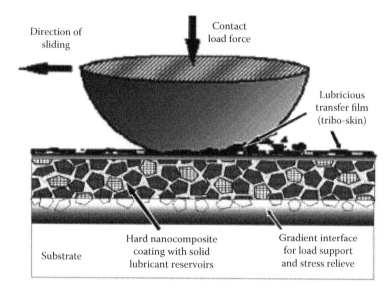

FIGURE 5.9 Schematic of the wear mechanism for self-lubricative coating. (From Voevodin, A.A. and Zabinski, J.S., *Comp. Sci. Technol.*, 65, 741–748, 2005.)

important to optimize the amount of reinforcement in the composites to achieve better tribological properties. Several other studies on self-lubricating composites with different lubricants have also shown the improved worn surface after tribological test [31–34].

The improvement in the worn surfaces is due to the formation of third bodies or solid lubricative films as a result of sliding on the self-lubricative composite surface. Thus, it reduces the friction and wear of the composites. It can be understood from Figure 5.9 that in the first step, the embedded solid lubricants come out from the matrix and form a lubricious film under the action of sliding motion and load. It leads to the second step where this lubricious film does not allow the contact between two surfaces and reduces friction.

Molecular dynamics simulation has not been reported for self-lubricating materials. However, the second step of wear mechanism suggests that the presence of the third body between the two rubbing surfaces can affect the tribological properties. In Section 5.3.2, we review the available molecular dynamics studies of the third body between two rubbing surfaces.

5.3.2 Effect of the Third Body between Two Rubbing Surfaces

Hu et al. investigated the effectiveness of Cu nanoparticle between two friction surfaces [35]. They prepared two models: model A without Cu nanoparticle and model B with Cu nanoparticle with two Fe blocks. Periodic boundary conditions were imposed in the x and z directions, whereas ends were fixed in the z direction as shown in Figure 5.10. The embedded atom method potential was used to describe the interaction between atoms (Fe–Fe, Cu–Cu, and Fe–Cu).

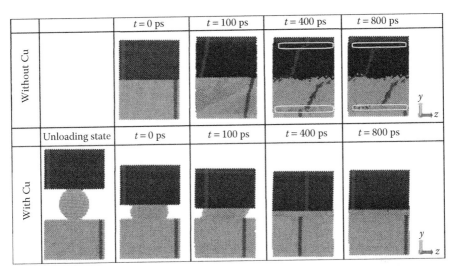

FIGURE 5.10 Frictional state during sliding at different times. Velocity = 10 m/s and load = 500 MPa. (From Hu, C. et al., *Appl. Surf. Sci.*, 321, 302–309, 2014.)

The molecular dynamics simulation was performed in LAMMPS. Initially, the systems were allowed to relax for 200 ps to achieve the equilibrium state. During sliding, frictional forces were monitored and different simulations were performed to test at various velocities and normal load. The deformation of Cu nanoparticle is presented in Figure 5.10. The blue and red markers were used to demonstrate the deformation of Fe blocks during sliding. The presence of Cu atoms reduced the friction between two Fe surfaces, which results in almost no deformation of the solid surface. The authors also reported that the anti-wear mechanism varies with the sliding velocity. At lower velocity (10 mm/s), a nano-film was formed, whereas at higher velocity (500 mm/s), transfer layers were formed due to diffusion of atoms.

In a similar study, Ewen investigated the effect of carbon nanoparticles between two Fe surfaces [36]. He considered two forms of carbon: carbon nanodiamonds and carbon nano-onions. The boundary conditions used in this study are the same as those in the previous study [35]. C–C interactions were described by adaptive intermolecular reactive empirical bond order potential. Fe–Fe interactions were described using embedded atom model potential and Lennard–Jones potential for Fe atoms in the same block and in opposite block, respectively. Lennard–Jones potential also modeled Fe-C van der Waals interaction. Simulated results are shown in Figures 5.11 and 5.12.

Figure 5.11 shows the effect of carbon particle coverage on the friction coefficient. The 1.00, 0.44, and 0.11 coverages were simulated by placing 9, 4, and 1 nanoparticles between Fe blocks, respectively. For higher coverage, the coefficient of friction is very low and zero wear, which is also manifested by the simulated image (Figure 5.12). The authors also investigated the effect of pressure (1–5 GPa) on friction properties, and they observed that at higher pressure and low coverage, the nanoparticle indented

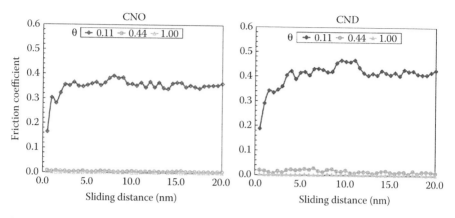

FIGURE 5.11   Effect of carbon particle coverage on the friction. Normal load = 1 GPa and sliding velocity = 10 m/s. (From Ewen, J.P. et al., *Tribol. Lett.*, 63, 38, 2016.)

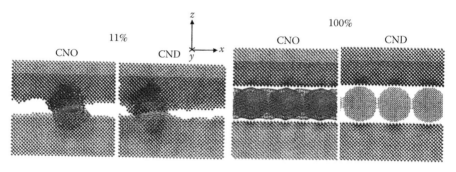

FIGURE 5.12   Simulated image of 0.11 and 1.00 coverage after 10 mm of sliding. Normal load = 1 GPa and sliding velocity = 10 m/s. (From Ewen, J.P. et al., *Tribol. Lett.*, 63, 38, 2016.)

and plowed the Fe block during sliding. Even under this condition, the friction was reduced by approximately 75% with respect to simulated study without nanoparticles; the obtained results were also matching with the experimental results.

Other than nanoparticles between surfaces, molecular dynamics have also been used to simulate the effect of adsorbed molecules on frictional surfaces [37,38]. The effect of the adsorbed stearic acid on α-Fe with different roughness and coverage was studied by Ewen et al. [38]. Simulations were performed on LAMMPS. First, the compression simulation was carried out for 0.5 ns followed by sliding simulation for 2.0 ns with +5 m/s velocity to the outermost layer of atoms in the top slab and −5 m/s velocity to the bottom slab. The variations of friction coefficient with surface coverage for different roughnesses are shown in Figure 5.13. The friction coefficient decreases with an increase in surface coverage for all rough surfaces.

Very few other studies have been carried out on the effect of the third body between two rubbing surfaces such as diamond and silicon dioxide confined by Fe blocks [39] and the effect of nanoparticles on the load-carrying capacity of lubricant

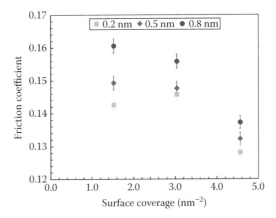

FIGURE 5.13 Variation in friction coefficient with surface coverage at 0.2, 0.5, and 0.8 nm root mean square (RMS) roughnesses. (From Ewen, J.P. et al., *Tribol. Inter.*, 107, 264–273, 2017.)

oil [40]. The obtained results are in good agreement with the experimental data. Thus, molecular dynamics simulation is capable of modeling or predicting the effect of the lubricious layer between two surfaces for different tribological parameters such as load, surface roughness, and coverage.

REFERENCES

1. Alder BJ, Wainwright TE. Phase transition for a hard sphere system. *The Journal of Chemical Physics*. 1957;27:1208–1209.
2. Multiscale molecular modeling, Molecular Simulation Engineering, accessed June 19, 2017. http://mose.units.it/Lists/Multiscale%20Molecular%20Modeling/AllItems.aspx.
3. Lees AW, Edwards SF. The computer study of transport processes under extreme conditions. *Journal of Physics C: Solid State Physics*. 1972;5:1921.
4. Gosling EM, McDonald IR, Singer K. On the calculation by molecular dynamics of the shear viscosity of a simple fluid. *Molecular Physics*. 1973;26:1475–1484.
5. Ashurst WT, Hoover WG. Argon shear viscosity via a Lennard-Jones potential with equilibrium and nonequilibrium molecular dynamics. *Physical Review Letters*. 1973;31:206.
6. Hoover WG. Nonequilibrium molecular dynamics. *Annual Review of Physical Chemistry*. 1983;34:103–127.
7. Harrison JA, Colton RJ, White CT, Brenner DW. Effect of atomic-scale surface roughness on friction: A molecular dynamics study of diamond surfaces. *Wear*. 1993;168:127–133.
8. Gromacs, accessed June 19, 2017. http://www.gromacs.org/.
9. NAMD - Scalable Molecular Dynamics, Theoretical and Computational Biophysics group, NIH Center for Macromolecular Modeling and Bioinformatics, Beckman Institute, University of Illinois, accessed June 19, 2017. http://www.ks.uiuc.edu/Research/namd/.
10. LAMMPS Molecular Dynamics Simulator, Sandia National Labs, accessed June 19, 2017. http://lammps.sandia.gov.
11. MedeA - Materials Design, accessed June 19, 2017. http://www.materialsdesign.com/medea.
12. Mate CM, McClelland GM, Erlandsson R, Chiang S. Atomic-scale friction of a tungsten tip on a graphite surface. *Physical Review Letters*. 1987;59:226–229.

13. Kim SH, Marmo C, Somorjai GA. Friction studies of hydrogel contact lenses using AFM: Non-crosslinked polymers of low friction at the surface. *Biomaterials*. 2001;22:3285–3294.

14. Ando Y. The effect of relative humidity on friction and pull-off forces measured on submicron-size asperity arrays. *Wear*. 2000;238:12–9.

15. Choi JS, Kim J-S, Byun I-S, Lee DH, Hwang IR, Park BH et al. Facile characterization of ripple domains on exfoliated graphene. *Review of Scientific Instruments*. 2012;83:073905.

16. Choi JS, Kim J-S, Byun I-S, Lee DH, Lee MJ, Park BH et al. Friction anisotropy–driven domain imaging on exfoliated monolayer graphene. *Science*. 2011;333:607–610.

17. Filleter T, Bennewitz R. Structural and frictional properties of graphene films on SiC (0001) studied by atomic force microscopy. *Physical Review B*. 2010;81:155412.

18. Mulliah D, Kenny SD, Smith R. Modeling of stick-slip phenomena using molecular dynamics. *Physical Review B*. 2004;69:205407.

19. Sørensen MR, Jacobsen KW, Stoltze P. Simulations of atomic-scale sliding friction. *Physical Review B Condensed Matter*. 1996;53:2101–2113.

20. Zhang L, Tanaka H. Towards a deeper understanding of wear and friction on the atomic scale—a molecular dynamics analysis. *Wear*. 1997;211:44–53.

21. Smith R, Mulliah D, Kenny SD, McGee E, Richter A, Gruner M. Stick slip and wear on metal surfaces. *Wear*. 2005;259:459–466.

22. Mulliah D, Kenny SD, McGee E, Smith R, Richter A, Wolf B. Atomistic modelling of ploughing friction in silver, iron and silicon. *Nanotechnology*. 2006;17:1807.

23. Liu XM, You X, Zhuang Z. Contact and Friction at Nanoscale. *Advanced Materials Research*. 2008;33–37:999–1004.

24. Chen L, Wang Y, Bu H, Chen Y. Simulations of the anisotropy of friction force between a silicon tip and a substrate at nanoscale. *Proceedings of the Institution of Mechanical Engineers, Part N: Journal of Nanoengineering and Nanosystems*. 2013;227:130–134.

25. Song J, Srolovitz DJ. Atomistic simulation of multicycle asperity contact. *Acta Materialia*. 2007;55:4759–4768.

26. Zhang H, Guo Z, Gao H, Chang T. Stiffness-dependent interlayer friction of graphene. *Carbon*. 2015;94:60–66.

27. Spijker P, Anciaux G, Molinari J-F. Relations between roughness, temperature and dry sliding friction at the atomic scale. *Tribology International*. 2013;59:222–229.

28. Bearings for vacuum environments, NSK Global, accessed June 19, 2017. http://www.nsk.com/products/spacea/vacuum/.

29. Hu KH, Wang J, Schraube S, Xu YF, Hu XG, Stengler R. Tribological properties of MoS_2 nano-balls as filler in polyoxymethylene-based composite layer of three-layer self-lubrication bearing materials. *Wear*. 2009;266:1198–1207.

30. Tabandeh-Khorshid M, Omrani E, Menezes PL, Rohatgi PK. Tribological performance of self-lubricating aluminum matrix nanocomposites: Role of graphene nanoplatelets. *Engineering Science and Technology, an International Journal*. 2016;19:463–469.

31. Voevodin AA, Zabinski JS. Nanocomposite and nanostructured tribological materials for space applications. *Composites Science and Technology*. 2005;65:741–748.

32. Moghadam AD, Omrani E, Menezes PL, Rohatgi PK. Mechanical and tribological properties of self-lubricating metal matrix nanocomposites reinforced by carbon nanotubes (CNTs) and graphene–A review. *Composites Part B: Engineering*. 2015;77:402–420.

33. Erdemir A. Review of engineered tribological interfaces for improved boundary lubrication. *Tribology International*. 2005;38:249–256.

34. Zhu S, Bi Q, Yang J, Liu W, Xue Q. Ni3Al matrix high temperature self-lubricating composites. *Tribology International*. 2011;44:445–453.

35. Hu C, Bai M, Lv J, Liu H, Li X. Molecular dynamics investigation of the effect of copper nanoparticle on the solid contact between friction surfaces. *Applied Surface Science*. 2014;321:302–309.

36. Ewen JP, Gattinoni C, Thakkar FM, Morgan N, Spikes HA, Dini D. Nonequilibrium molecular dynamics investigation of the reduction in friction and wear by carbon nanoparticles between iron surfaces. *Tribology Letters*. 2016;63:38.

37. He G, Robbins MO. Simulations of the static friction due to adsorbed molecules. *Physical Review B*. 2001;64:035413.

38. Ewen JP, Restrepo SE, Morgan N, Dini D. Nonequilibrium molecular dynamics simulations of stearic acid adsorbed on iron surfaces with nanoscale roughness. *Tribology International*. 2017;107:264–273.

39. Hu C, Bai M, Lv J, Kou Z, Li X. Molecular dynamics simulation on the tribology properties of two hard nanoparticles (diamond and silicon dioxide) confined by two iron blocks. *Tribology International*. 2015;90:297–305.

40. Hu C, Bai M, Lv J, Li X. Molecular dynamics simulation of mechanism of nanoparticle in improving load-carrying capacity of lubricant film. *Computational Materials Science*. 2015;109:97–103.

Index

Note: Page numbers followed by f and t refer to figures and tables respectively.